Concise Textbook of Equine Clinical Practice Book 4

This concise, practical text covers the essential information veterinary students need to succeed in equine medicine and surgery, focussing on skin, urinary, liver and endocrine diseases. Written for an international readership, the book conveys the core information in an easily digestible, precise form with extensive use of bullet points, tables, flow charts, diagrams, lists, protocols and extensive illustrations.

Part of a five-book series that extracts and updates key information from Munroe's *Equine Surgery, Reproduction and Medicine,* Second Edition, the book distils best practice in a logical straightforward clinical-based approach. It details clinical anatomy, physical clinical examination techniques, diagnostic techniques and normal parameters, emphasising the things regularly available to general practitioners with minimal information of advanced techniques.

- The liver section is divided into clinical evaluation, diagnostic tests, possible causes and treatment.
- The diagnostic approach to endocrine disease is followed by specific diseases of the thyroid, parathyroid, pituitary gland, equine metabolic syndrome and other miscellaneous disorders.
- The approach to urinary tract problems includes information on diagnostic tests and imaging, renal diseases, and diseases of the ureter, bladder and urethra.
- The approach to skin disease is discussed in detail, including relevant diagnostic tests. Following this, individual skin diseases are covered under the headings of genetic, infectious, immune-mediated, nutritional, endocrine, idiopathic, traumatic and neoplastic causes.
- Finally, skin wounds, burns and infections of synovial structures are discussed.

Ideal for veterinary students and nurses on clinical placements with horses as well as practitioners needing a quick reference 'on the ground'.

Concise Textbook of Equine Clinical Practice Book 4

Liver, Endocrine, Urinary, Skin and Wounds

François-René Bertin

Antonio Cruz

Andy Durham

Derek Knottenbelt

Edited By

Graham Munroe

CRC Press
Taylor & Francis Group
Boca Raton London New York

CRC Press is an imprint of the
Taylor & Francis Group, an **informa** business

First edition published 2024
by CRC Press
2385 NW Executive Center Drive, Suite 320, Boca Raton, FL 33431

and by CRC Press
4 Park Square, Milton Park, Abingdon, Oxon, OX14 4RN

CRC Press is an imprint of Taylor & Francis Group, LLC

ISBN: 9781032592268 (hbk)
ISBN: 9781032066172 (pbk)
ISBN: 9781003453666 (ebk)

DOI: 10.1201/9781003453666

Typeset in Sabon
by Evolution Design & Digital Ltd (Kent)

Printed in the UK by Severn, Gloucester on responsibly sourced paper

Table of Contents

Preface

A vast array of clinical equine veterinary information is available for the under- and post-graduate veterinarian and veterinary nurse to peruse. This is contained in textbooks, both general and specialised, and increasingly online at websites of varying quality and trust-worthiness. It is easy for the veterinary student or nurse, recent graduate and busy general or equine practitioner to become overwhelmed and confused by this diverse range of information. Often what is required, particularly in the clinical situation, is a distillation of the essential knowledge and best practice required to treat a horse in the most suitable way. This concise, practical text is designed to provide the essential information needed to understand and treat clinical cases in equine practice.

This book focuses on medical and surgical conditions of the liver, endocrine system, urinary tract, skin and wounds. It is part of a five-book series, which between them will cover all the areas of equine clinical practice. The information is extracted and updated from *Equine Clinical Medicine, Surgery and Reproduction* (Second Edition), which was published in 2020. It is written for an international readership and is designed to convey the core, best-practice information in an easily digested, quick reference form using bullet points, lists, tables, flow charts, diagrams, protocols and extensive illustrations and photographs.

The liver section is divided into clinical evaluation, diagnostic tests, possible causes and treatment. The diagnostic approach to endocrine disease is followed by specific diseases of the thyroid, parathyroid, pituitary gland, equine metabolic syndrome and other miscellaneous disorders. The approach to urinary tract problems includes information on diagnostic tests and imaging, renal diseases, and diseases of the ureter, bladder and urethra. The important approach to skin disease in the horse is discussed in some detail, including relevant diagnostic tests. Following this, individual skin diseases are covered under the headings of genetic, infectious, immune-mediated, nutritional, endocrine, idiopathic, traumatic and neoplastic causes. Finally, skin wounds, burns and infections of synovial structures are discussed. All the material is approached in the same logical, straightforward, clinical-based way. There are details of relevant clinical anatomy, physical clinical examination techniques, normal parameters, aetiology/pathophysiology, clinical examination findings, differential diagnosis, diagnostic techniques, management and treatment and prognosis. The emphasis is on information tailored to general equine clinicians with just enough on advanced techniques to make the practitioner aware of what is available elsewhere.

The intention of this series of books is for them to be used on a day-to-day basis in clinical practice by student and graduate veterinarians and nurses. The spiral binding format allows them to lie open on a surface near to the patient, readily available to the veterinary student or practitioner whilst looking at, or treating, a clinical case.

About the Authors

François-René Bertin graduated with a DVM from the National Veterinary School of Nantes (France) and completed an internship at the National Veterinary School of Alfort (France). He trained in Equine Internal Medicine at Purdue University (USA) and became a Diplomate of the American College of Veterinary Internal Medicine (ACVIM). Dr Bertin completed his PhD at McGill University (Canada). He joined The University of Queensland (UQ) in 2016 and has authored several research articles, book chapters and the first textbook on the diagnosis and management of equine endocrinopathies. He leads the Equine Endocrinology research group at UQ and is a member of international expert panels to create guidelines for the management of insulin dysregulation and PPID.

Antonio Cruz is double-boarded in Equine Surgery and Sports Medicine. He is currently a faculty member at the Justus Liebig Universität Giessen in Germany having spent most of his career in North America. He has spent 25 years in academic practice in different capacities, having been tenured faculty (Associate Professor) at the University of Guelph (Canada) for several years. He has also worked at the Universities of Saskatchewan, Minnesota, Prince Edward Island and Bern in Switzerland. He also spent 7 years in private practice establishing a leading referral surgical and sports medicine facility in Vancouver, Canada. He holds several post-graduate degrees. He has supervised many graduate students, residents and interns and published over 60 articles and many book chapters. He is a regular speaker at international meetings. His main clinical focus is equine orthopaedics, and he is actively involved in research in the areas of gait analysis and equine surgery.

Andy Durham graduated in 1988 and has worked in private practice throughout his entire career. After 2 years in mixed practice, Dr Durham entered equine practice in 1991. He joined the Liphook Equine Hospital in 1994 where he still works as Clinical Director. Dr Durham gained the RCVS Certificate in Equine Practice in 1993, the RCVS Diploma in Equine Internal Medicine in 2003 and Diplomate of the European College of Equine Internal Medicine (DECEIM) in 2004. He is an RCVS and European Specialist in Equine Internal Medicine and Visiting Professor at the University of Surrey. His clinical interests span a wide range of equine internal medical problems including endocrinopathies, abdominal disease, nutrition, liver disease, ophthalmology, dermatology, critical care and infectious diseases. His daily work is split between equine medicine referral work and the clinical diagnostic laboratory at Liphook.

Derek Knottenbelt qualified in Edinburgh and retired from his personal Chair in Equine Internal Medicine at the University of Liverpool in 2010. He is a Diplomate of the ECEIM and the ACVIM. He is Director of Equine Medical Solutions Ltd and through this provides an advisory service in equine oncology. He has been awarded honorary life membership of ECEIM, BEVA, VWHA and ACVIM. He received the Animal Health Trust Scientific Award in 2003, the BEVA (Blue Cross) Welfare Award in 2004 and the Merck–WEVA Welfare Award in 2006. In 2005, he was appointed OBE by the Queen for his services to equine medicine. He has published widely on a range of topics and maintains an interest in equine internal medicine and oncolog, in particular. His main interests are in oncology, ophthalmology, wound management and dermatology.

Graham Munroe qualified from the University of Bristol with honours in 1979. He spent 9 years in equine practice in Wendover, Newmarket, Arundel and Oxfordshire, and a stud season in New Zealand. He gained a certificate in equine orthopaedics and a diploma in equine stud medicine from the RCVS whilst in practice. He joined Glasgow University Veterinary School in 1988 as a lecturer and then moved to Edinburgh Veterinary School as a Senior Lecturer in Large Animal Surgery from 1994 to 1997. He obtained by examination a Fellowship of the RCVS in 1994 and Diplomate of the ECVS in 1997.

He was awarded a PhD in 1994 for a study in neonatal ophthalmology. He has been a visiting equine surgeon at the University of Cambridge Veterinary School, University of Bristol Veterinary School and Helsingborg Hospital, Sweden. He was Team Veterinary Surgeon for British Driving Teams, 1994–2001; British Dressage Teams, 2001–2002; and the British Vaulting Team in 2002. He was FEI Veterinary Delegate at the Athens 2004 Olympics. He currently works in private referral surgical practice, mainly in orthopaedics. He has published over 60 papers and book chapters.

Abbreviations

AAAs	Aromatic amino acids
acetyl-CoA	Acetyl coenzyme A
ACTH	Adrenocorticotropic hormone
AKI	Acute kidney failure
ALP	Alkaline phosphatase
α-MSH	α-Melanocyte-stimulating hormone
ANA	Antinuclear antibody testing
APTT	Activated partial thromboplastin time
ASIT	Allergen-specific immunotherapy
AST	Aspartate aminotransferase
BCAAs	Branched-chain amino acids
BCG	Bacillus Calmette–Guérin therapy
BCS	Body condition score
BH	Biliary hyperplasia
BPV	Bovine papillomavirus
CBC	Complete blood count
CE	Cystic echinococcosis
CFT	Complement fixation test
CIRCI	Critical illness-related corticosteroid insufficiency
CKD	Chronic kidney disease
CLE	Equine cutaneous lupus erythematosus
CLIP	Corticotropin-like intermediate peptide
CRT	Capillary refill time
CT	Computed tomography
CTCL	Cutaneous T-cell lymphoma
DTM	Dermatophyte test medium
EDNRB	Endothelin receptor type B
EGT	Exuberant granulation tissue management
EHV	Equine herpesvirus
ELISA	Enzyme-linked immunosorbent assay
EMS	Equine metabolic syndrome
EPM	Equine protozoal meningitis/myeloencephalopathy
EZL	Epizootic lymphangitis
F	Fibrosis
FFAs	Free fatty acids
GCT	Granulosa (thecal) cell tumour
GFR	Glomerular filtration rate
GGT	Gamma-glutamyltransferase
GI	Gastrointestinal
GLDH	Glutamate dehydrogenase
HE	Hepatic encephalopahy
ICSs	Intercostal spaces
IDH	Iditol dehydrogenase
IFA	Immunofluorescent antibody
IHC	Immunohistochemistry

KOH	Potassium hydroxide
LA	Local anaesthetic
LDH	Lactate dehydrogenase
LE	Lupus erythematosus
LMN	Lower motor neuron
LOLA	L-ornithine-L-aspartate
M	Megalocytosis
MRI	Magnetic resonance imaging
NEFAs	Non-esterified fatty acids
NSAIDs	Non-steroidal anti-inflammatory drug
PCLCV	Pastern and canon leukocytoclastic vasculitis
PCR	Polymerase chain reaction
PDT	Photodynamic therapy
PPID	Pituitary pars intermedia dysfunction
PRP	Platelet-rich plasma
PT	Prothrombin time
PU/PD	Polyuria/polydipsia
qPCR	Real-time polymerase chain reaction
RAI	Relative adrenal insufficiency
RSM	Rapid sporulating medium
RT	Reverse transcription
RTA	Renal tubular acidosis
SAA	Serum amyloid A
SAMe	S-adenyl methionine (SAMe)
SDH	Sorbitol dehydrogenase
SG	Urine-specific gravity
SGLT2	Sodium glucose co-transport inhibitors
SLE	Systemic lupus erythematosus
T_4	Thyroxin
TCRLBCL	T-cell-rich, large B-cell cutaneous lymphoma
TNP	Topical negative pressure
TRH	Thyrotropin-releasing hormone
TSH	Thyroid-stimulating hormone
UFE	Urinary fractional excretion
UMN	Upper motor neuron
UP/UC	Urine protein:creatinine ratio
VLDLs	Very-low-density lipoproteins
WBCs	White blood cells

Liver Disease

CLINICAL EVALUATION OF HORSES WITH LIVER DISEASE

- Liver has a large functional reserve:
  - considerable damage (> 60–70%) before normal metabolic functions significantly compromised.
- acute hepatic insufficiency less common than slow/insidious progressive loss of hepatic function.
- hepatic insufficiency may lead to gradually progressive appearance of clinical signs:
  - weight loss and lethargy.
  - acute decompensation may occur where previously there has been a slow and gradual development of hepatic damage.
  - signs of hepatic encephalopathy:
    - ◆ severe cerebral dysfunction (depression, circling, head-pressing, blindness).
    - ◆ bilateral laryngeal paralysis may develop acutely.
- functional effect of liver disease is important:
  - horses may have liver damage that does not interfere significantly with hepatic metabolic activities.
  - other horses have liver damage which results in failure of hepatic function:
    - ◆ clinical signs usually only develop at this point.
    - ◆ high likelihood of more severe hepatic pathology.
- hepatic insufficiency clinical signs (Figs. 1.1–1.5) include:
  - lethargy
  - inappetence
  - weight loss
  - photodermatitis
  - oedema
  - central and peripheral neurological dysfunction
  - jaundice
  - polydipsia
  - pruritus
  - bleeding
  - colic
- many horses with mild/moderate liver disease remain outwardly healthy:
  - adequate hepatic function maintained (**subclinical hepatopathy**).
  - absence of clinical signs tells you nothing about the health of the liver.

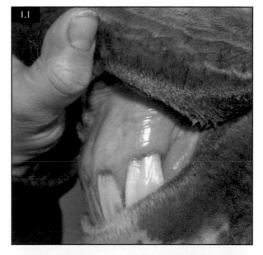

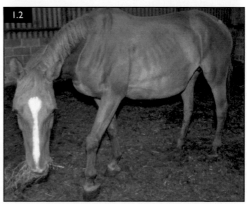

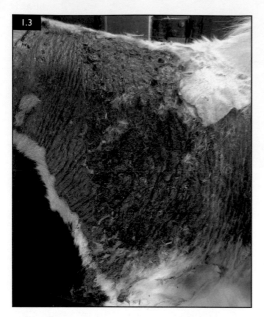

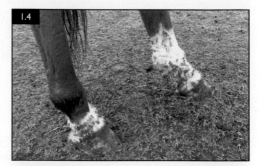

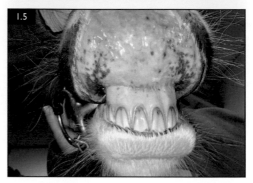

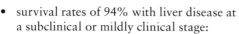

FIGS. 1.1, 1.2, 1.3, 1.4, 1.5 Images of several clinical signs of hepatic failure, comprising jaundice (1.1), weight loss and ventral oedema (1.2), photodermatitis (1.3, 1.4) and petechiation (1.5).

- survival rates of 94% with liver disease at a subclinical or mildly clinical stage:
  - 49% surviving when clinical signs of hepatic insufficiency are present.
  - early investigation of liver disease is advisable.

Hepatic encephalopathy (HE)

- may result from hepatic insufficiency or, rarely, portosystemic bypass.
- may present with:
  - obtundation
  - behavioural changes
  - yawning (Fig. 1.6)
  - disorientation
  - compulsive walking
  - ataxia
  - circling
  - bilateral laryngeal paralysis (Fig. 1.8)
  - blindness
  - head-pressing
  - ptyalism (Fig. 1.7)
  - seizure
  - coma
  - pruritus
  - dysphagia
  - gastric impaction
  - foot stamping
- poorer prognosis than liver disease without HE but attempted treatment is justifiable.
- many pathophysiological factors may contribute to HE, including:
  - ammonia
  - tumour necrosis factor-alpha
  - aromatic amino acids
  - manganese
  - copper
  - phenols
  - benzodiazepine-like substances
  - mercaptans
  - short-chain fatty acids
  - monoamines
  - neurosteroids
  - bilirubin
  - electrolytes
- ammonia plays a major role:
  - nearly all HE cases have high plasma ammonia.
  - poor correlation between plasma ammonia concentration and severity of HE.
  - altered blood–brain barrier permeability to ammonia may play a role.

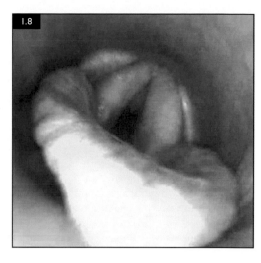

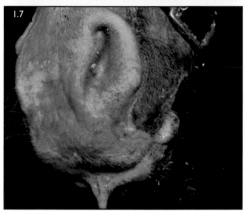

FIGS. 1.6, 1.7, 1.8 Images demonstrating signs of hepatic encephalopathy, including somnolence and head-pressing (1.6), ptyalism (1.7) and respiratory distress associated with bilateral laryngeal paralysis (1.8).

- astrocytes are sole source of glutamine synthetase in the brain:
  - catalyses conversion of cerebral ammonia to glutamine.
- osmotic effect of synthesised glutamine causes astrocyte cell swelling:
  - acute HE characterised by overt brain oedema.
  - chronic HE associated with milder oedema, astrocyte swelling and dysfunction.

Photosensitisation

- hepatopathy should be considered in all photodermatitis cases.
- phylloerythrin is a photodynamic agent absorbed from the colon following degradation of chlorophyll.
- hepatogenous (or secondary) photosensitisation results from failure of the liver to detoxify and excrete phylloerythrin.
- increased circulating phylloerythrin concentrations and consequent photoactivation in superficial blood vessels of non-pigmented skin areas lead to skin injury.

- pruritus may be an early feature, followed by pain as the condition progresses.
- typical signs (see Figs. 1.3, 1.4) are:
  - erythema of exposed non-pigmented skin.
  - progression to serous exudation, local oedema, crusting and necrosis.
- may be seen in the absence of other clinical/clinicopathological signs of hepatic insufficiency.

Jaundice

- sequential enzymatic conversion of haemoglobin to haem to biliverdin and bilirubin:
  - during erythrocyte degradation and normal turnover or haemolysis.
- unconjugated (indirect) bilirubin:
  - normally extracted from the circulation by hepatocytes.
  - conjugate the molecule (direct bilirubin) before excretion into bile.
- systemic accumulation of bilirubin leads to jaundice (see Fig. 1.1):
  - some normal horses appear slightly jaundiced.

- o retinoid accumulation but normal plasma bilirubin levels.
- jaundice develops in hepatic insufficiency due to:
  - o failure to clear unconjugated (indirect) bilirubin from plasma and/or
  - o failure to excrete conjugated (direct) bilirubin following biliary stasis.
  - o majority of bilirubin in plasma will be unconjugated even in biliary stasis.
- other causes of jaundice with hyperbilirubinaemia are:
  - o anorexia/hypophagia (common).
  - o hacmolytic anaemia (uncommon).
  - o Gilbert's syndrome (rare genetic failure to process bilirubin)

Bleeding and clotting disorders

- hepatic failure can lead to impaired coagulation causing prolonged:
  - o activated partial thromboplastin time (APTT) and prothrombin time (PT).
- vitamin K-dependent factors are generally affected first:
  - o prothrombin, factors VII, IX and X and proteins C and S.
- many further factors and platelets eventually affected by hepatic insufficiency.
- laboratory measurements of coagulation are commonly abnormal in hepatic failure:
  - o clinical evidence of bleeding is less common:

- ♦ petechiation (see Fig. 1.5) and echymoses.
- ♦ epistaxis or prolonged bleeding after venepuncture.
- no evidence that abnormal clotting tests increase the risk of liver biopsy:
  - o checking clotting times pre-biopsy is not necessary.

Diagnostic blood tests in liver disease

- following clinical examination, analysis of blood samples should be carried out.
- useful serum biochemical tests (Table 1.1) are divisible into:
  - o intracellular enzymes released into the circulation following damaged liver cells.
  - o other biochemical substances that reflect the effectiveness of various hepatic functions.

Hepatic enzymes

- high diagnostic value in the evaluation of suspected hepatic insult:
  - o no enzyme is entirely specific for liver disease:
    - ♦ may not be raised in every case of liver disease.
  - o useful along with other diagnostic data in raising suspicion of liver disease in individuals or groups.
  - o high serum concentrations of liver enzymes may be associated in some cases with a poorer prognosis.

TABLE 1.1 Biochemical substances measurable in serum and plasma that may reflect hepatic injury and function	
LIVER-DERIVED ENZYMES	**BIOCHEMICAL MARKERS OF HEPATIC FUNCTION**
<ul><li>Alkaline phosphatase (ALP)</li><li>Aspartate aminotransferase (AST)</li><li>Gamma-glutamyltransferase (GGT)</li><li>Glutamate dehydrogenase (GLDH)</li><li>Iditol dehydrogenase (IDH)</li><li>Lactate dehydrogenase (LDH)</li><li>Sorbitol dehydrogenase (SDH)</li></ul>	<ul><li>Acute phase proteins (serum amyloid A, fibrinogen)</li><li>Albumin</li><li>Amino acids</li><li>Ammonia</li><li>Bile acids</li><li>Bilirubin (total, unconjugated, conjugated)</li><li>Clotting tests (APTT, PT)</li><li>Creatinine</li><li>Globulins</li><li>Glucose</li><li>Urea</li></ul>

- aspartate aminotransferase (AST) and lactate dehydrogenase (LDH):
  - may come from multiple tissue types:
    - liver and muscle are main origin of increased serum concentrations.
- serum gamma-glutamyltransferase (GGT), glutamate dehydrogenase (GLDH) and sorbitol dehydrogenase (SDH):
  - widely regarded as liver specific but lacks an evidence basis in horses.
  - other tissues, including the gastrointestinal tract, may contribute in some cases.
  - mild/moderately increased liver enzymes in some primary gastrointestinal (GI) diseases:
    - colon impaction and displacement, and gastric ulceration.
- serum biochemical indicators of liver disease are found in clinically healthy horses:
  - subclinical liver disease is not unusual.
- serum biochemistry useful in establishing an outbreak of liver disease:
  - following identification of a horse with evidence of liver disease:
    - horses on same premises evaluated using biochemistry (even if appear healthy).
    - widens case base and directs epidemiological considerations towards possible infectious or nutritional/toxic causes.

Markers of hepatic function/dysfunction

- more prognostic information than serum enzymes but not straightforward.
- unconjugated bilirubin and bile acids normally removed from the circulation, by functioning hepatocytes:
  - accumulation in serum indicative of failure of hepatocyte function.
  - both are increased in a catabolic state in the absence of hepatic disease:
    - fasting, anorexia or hypophagia.
- total serum globulins often increased in horses with hepatic failure:
  - loss of hepatic macrophagic Kupffer cells (line liver sinusoids) may stimulate generalised immune reaction.
  - possibly a stronger prognostic factor than any other functional analyte.

- hepatic insufficiency sometimes associated with hypoalbuminaemia in horses:
  - less common and milder effect than in other species.
  - probably due to a longer half-life for albumin in the horse.
- serum urea and creatinine may be abnormally low in hepatic failure:
  - low urea may arise from failure of ureagenesis in the liver.
  - urea and creatinine may be decreased by polyuria following failure of hepatic aldosterone degradation.
- ammonia produced as normal part of protein catabolism is not detoxified in liver failure:
  - high plasma ammonia useful to confirm HE.
  - ammonia concentrations are very unstable in vitro:
    - samples must be chilled promptly.
    - analysed within 4 hours of collection.
- hypoglycaemia rarely seen in horses with hepatic failure:
  - more likely in severely ill cases probably because of insulin resistance.
  - hepatic neoplasia with profound intermittent hypoglycaemia rarely reported.
- abnormal serum concentrations of amino acids commonly associated with hepatic insufficiency:
  - increased methionine and aromatic amino acids (AAAs), tyrosine and phenylalanine.
  - decreased branched-chain amino acids (BCAAs) valine, leucine and isoleucine.
  - rarely offered as tests by commercial laboratories.

Other blood tests

- haematology rarely gives specific diagnostic help in liver disease cases:
  - relative neutropaenia and lymphocytosis sometimes observed.
  - erythrocytosis:
    - may reflect dehydration, especially in sick cases.
    - some neoplastic/non-neoplastic hepatopathy cases without dehydration.
- serum markers are available for indicating the degree of hepatic fibrosis:

- o fibrosis is single most useful prognostic indicator in hepatic disease.
- o very useful when liver biopsy is not performed.

Ultrasonographic assessment of the liver

- hepatic tissue is typically only about 3 cm deep to the skin surface:
  - o easily imaged by ultrasonography.
  - o majority of liver masked by echo-reflective lungs and cannot be imaged.
- primary use is to guide liver biopsy.
- imaged via both right and left hemithoraces in most horses:
  - o more liver usually imaged on the right than left:
    - ♦ occasionally, reverse is true or imaged only on the left.
    - ♦ occasional cases where no liver tissue imaged by ultrasonography.
- normal hepatic image from the right side:
  - o approximately triangular in shape.
  - o sharply angled caudoventral edge (Fig. 1.9).
  - o convex surface adjacent to the diaphragm laterally.
  - o concave surface against the hyperechoic colonic image medially.
- normal hepatic image from the left side:
  - o biconvex shape is often seen (Fig. 1.10).
  - o do not confuse with the caudomedially adjacent spleen:
    - ♦ similar appearance but more hyperechoic and fewer blood vessels.
- normal liver relatively hypoechoic homogeneity, interrupted by anechoic blood vessels:
  - o bile ducts cannot normally be visualised.
  - o larger blood vessels:
    - ♦ flow often visualised in real time with standard B-mode ultrasonography.
    - ♦ Doppler will establish active blood flow where doubt exists over identity of dilated vessels (Fig. 1.11).
  - o blood vessels are typically <9 mm in diameter in the superficial regions:
    - ♦ several dilated (>9 mm) blood vessels in the superficial hepatic

areas may be associated with portal hypertension and hepatic fibrosis.
  - o larger blood vessels sometimes seen deeper in the liver in ponies due to the ability to image at that level:
    - ♦ large central caudal vena cava often seen from the right side in ponies and should not be confused with a cystic structure.
  - o very localised hyperechoic (without acoustic shadows) signals are commonly associated with blood vessels:
    - ♦ fibrous perivascular connective tissue and acoustic enhancement artefact.
- most reliable way to locate the liver by ultrasonography is:
  - o left side:
    - ♦ liver typically imaged over 2–4 intercostal spaces between 5th and 9th (see Fig. 1.9).
    - ♦ immediately caudal to heart and cranial to spleen and stomach.
  - o right side:
    - ♦ usually (not always), larger image area (three and eight intercostal spaces).
    - ♦ typically, anywhere between 6th and 16th intercostal spaces (see Fig. 1.10).
  - o place ultrasound transducer over the thoracic lung field (left or right):
    - ♦ obtain characteristic and easily recognisable echo-reflective gliding pleural surface with reverberation artefact.
  - o move the transducer in a ventral direction:
    - ♦ liver image will become visible immediately ventral to the border of the lung.
- **most cases of hepatopathy do not have ultrasonographic abnormalities.**
- images classified as abnormal:
  - o high specificity for presence of significant liver disease.
  - o associated with poorer outcomes:
  - o fibrosis, haemosiderosis and lipidosis associated with diffuse increase in echogenicity of hepatic tissue:
    - ♦ subjective.
    - ♦ compare with splenic echogenicity:

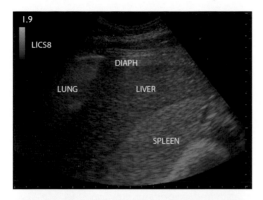

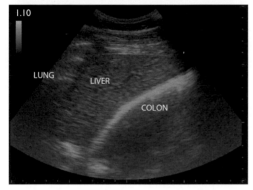

FIGS. 1.9, 1.10 Ultrasonographic images of the normal liver as seen via the left 8th intercostal space (1.9) and the right 12th intercostal space (1.10).

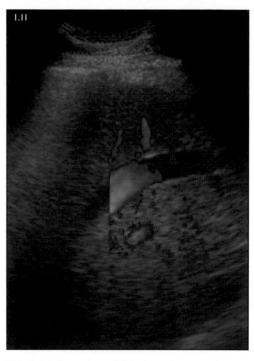

FIG. 1.11 An enlarged intrahepatic vessel confirmed to be a blood vessel by active flow detected using colour flow Doppler.

- spleen normally more echogenic than the liver on left side.
- hepatomegaly or atrophy is subjective:
  ○ area of liver imaged depends on size of lung fields and size of liver *per se*.
  ○ approximately 10–12 cm of hepatic tissue imaged projecting caudoventrally from expiratory border of lung in right 13th intercostal space.
  ○ peripheral margins of the liver should demonstrate acute angulation:
    ♦ smooth or rounded supports hepatomegaly and swelling.
  ○ occasionally no hepatic tissue seen at all on right side of normal horses.
- single or multifocal small (<5 mm) hyperechoic foci occasionally encountered:
  ○ absence of other changes.
  ○ 'Starry sky' pattern (Fig. 1.12).
  ○ rarely pathologically significant.

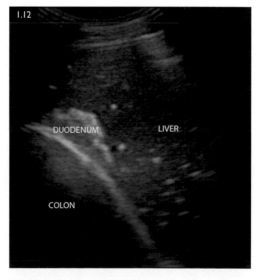

FIG. 1.12 The so-called 'starry sky' pattern of multifocal hepatic echogenicity in a horse with no other evidence of liver disease.

Liver biopsy

- very valuable in the management of equine liver disease:
  - establish a diagnosis.
  - establish prognosis – most accurate means.
  - guide selection of appropriate therapy by determining pattern of pathology:
    - aetiology may sometimes be revealed.
- safe procedure:
  - especially under ultrasound guidance
  - caution advised if hepatic insufficiency present:
    - higher risks of procedure.
    - additional clotting factors with a fresh plasma transfusion pre-biopsy.
    - no evidence to support the use of pre-biopsy evaluation of haemostasis.
- suitable biopsy site(s) within the liver is/ are chosen with ultrasonography based on:
  - thickness of imaged hepatic tissue.
  - absence of large blood vessels.
  - occasionally, focal appearance of abnormal hepatic tissue.
  - external landmarks cannot be used to reliably identify a suitable site.
- biopsy needles 14 gauge and preferably at least 10–15 cm long:
  - three main types:
    - simple hand-operated 'Tru-Cut' type needle (Fig. 1.13).
    - semi-automatic spring-loaded needle.
    - larger/more powerful biopsy-gun device with simple needle.
- most biopsies collected from the right side due to:
  - larger hepatic mass.
  - avoidance of proximity to the left ventricle.

Procedure

- light sedation (generally an alpha-2-antagonist plus butorphanol).
- select suitable site or sites.
- clip and aseptic preparation.
- single dose of a non-steroidal anti-inflammatory drug (NSAID) to minimise possible post-biopsy discomfort.
- single dose of Procaine penicillin.
- 5 ml local anaesthetic solution subcutaneously (e.g. mepivacaine).
- small stab incision in the skin (#11 or #15 scalpel blade).
- insert biopsy needle to predetermined depth and angle based on ultrasonographic image:
  - most cases, needle introduced between 4 and 6 cm deep to skin surface.
- slight cranial orientation of needle allows deeper target of hepatic tissue (Fig. 1.14):
  - spring-loaded biopsy needles function best inserted at 90 degrees to skin surface.
- biopsy can be collected under real-time ultrasound guidance:
  - preferably with a biopsy guide attached to the transducer.

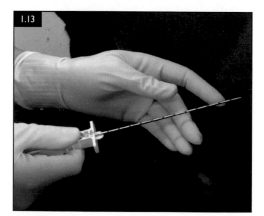

FIG. 1.13 A liver biopsy being collected using a hand-operated Tru-Cut needle.

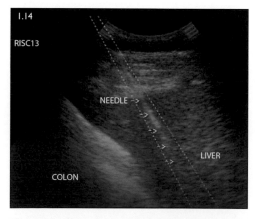

FIG. 1.14 Ultrasonographic image showing the biopsy needle within the liver during the biopsy procedure.

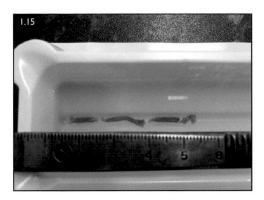

FIG. 1.15 Liver biopsy specimens of mixed individual lengths collected from the same horse (approximately 33 mm of biopsy specimens in total).

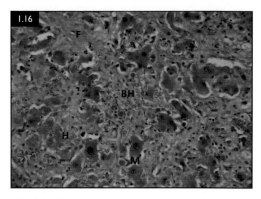

FIG. 1.16 Biopsy specimen demonstrating typical signs of pyrrolizidine alkaloid toxicosis including fibrosis (F), megalocytosis (M) and biliary hyperplasia (BH), with an additional finding of orange granular haemosiderosis (H).

- more simply, based on depth and angle measurements taken from still ultrasound images at site of biopsy.
- when biopsy needle in desired position, activate to cut the biopsy before withdrawal.
- multiple biopsies are required:
  ○ biopsy specimens do not always completely fill the biopsy chamber.
  ○ approximately 2–4 cm of biopsy material collected in total (Fig. 1.15).
- specimens placed in 10% buffered formalin for histopathology.
- bacterial cholangiohepatitis suspected:
  ○ useful to culture fresh biopsy material.
  ○ place in sterile container with small amount of saline.

- prognostic liver biopsy scoring system is reasonably accurate:
  ○ score based on broad comparative index of histopathological severity.
  ○ weighted scoring in five different pathological findings (Table 1.2):
    ◆ fibrosis.
    ◆ biliary hyperplasia.
    ◆ inflammation.
    ◆ haemosiderosis (within hepatocytes or Kupffer cells).
    ◆ irreversible cytopathology (necrosis and/or megalocytosis).
- total biopsy score may range from 0 to 14:
  ○ good prognosis associated with scores < 3.
  ○ very poor prognosis with scores > 7.

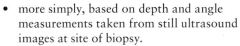

TABLE 1.2 Biopsy scoring system

VARIABLE	ABSENT	MILD	MODERATE	SEVERE
Fibrosis	0	0	2	4
Irreversible cytopathology	0	1	2	2
Inflammatory infiltrate	0	0	1	2
Haemosiderin accumulation	0	0	0	2
Biliary hyperplasia	0	0	2	4

Scores attributed to the presence and severity of five pathological changes.
Minimum score = 0, maximum score = 14.
Irreversible cytopathology = necrosis or megalocytosis.
(From Durham *et al*. 2003a.)

POSSIBLE CAUSES OF LIVER DISEASE

- identification of the actual cause of liver disease is difficult:
  - uncommonly identified on basis of clinical signs, blood tests or ultrasound.
- conditions where aetiology can be inferred based on biopsy findings include:
  - septic cholangiohepatitis.
  - pyrrolizidine alkaloid toxicosis (Fig. 1.16).
  - neoplasia.
  - severe haemosiderosis.
- majority of liver biopsy specimens have patterns of histopathology that are not definitive for causation.

Hepatoxicity

- hepatic disease commonly occurs as an outbreak in equids:
  - index case is usually first identified as part of a clinical investigation.
  - further herd mates with subclinical hepatopathy then usually identified by speculative testing.
  - most hepatotoxins do not lead to specific findings in liver biopsy specimens:
    - prime exception is pyrrolizidine alkaloids (Pas):
      - megalocytosis, biliary hyperplasia and fibroplasia are typical.
  - pattern of multiple simultaneous cases suggestive of hepatotoxicity or infectious disease.
- several hundred plant species are known to be hepatotoxic to the horse:
  - pyrrolizidine alkaloid-containing species:
    - *Senecio, Cynoglossum, Echium* and *Crotalaria*.
    - *Amsinckia, Symphytum, Chimonanthus* and *Heliotropium*.
  - kleingrass (*Panicum coloratum*)
  - cocklebur (*Xanthium* sp.)
  - *Cestrum* spp.
  - *Senna* spp.
  - *Indigofera* spp.
  - *Lantana* spp.

- red and alsike clover (*Trifolium pratense* and *T. hybridum*):
  - probably mycotoxins associated with the plants.
- many drugs used in equine medicine have potential for hepatotoxicity including:
  - amiodarone
  - anabolic steroids
  - azathioprine
  - carbamazepine
  - cyproheptadine
  - erythromycin
  - isoniazid
  - NSAIDs
  - omeprazole
  - paracetamol
  - penicillin
  - phenobarbitone
  - phenothiazines
  - phenytoin
  - rifampin
  - sulphonamides
  - tetracyclines
- iron is accumulated in the liver of normal domesticated horses:
  - mild to moderate amounts of hepatic haemosiderin are common in horses without evidence of liver disease.
  - interpretation of the relevance of large amounts of haemosiderin in the liver should consider:
    - further signs of pathology.
    - age of animal.
    - relative accumulation in hepatocytes versus Kupffer cells and other macrophages.
  - iron accumulation and toxicity may be associated with:
    - iron-rich supplements in an acute or chronic fashion.
    - some forages are high in iron, usually when soil included in the forage during grass cutting,
    - some natural water sources can be high in iron oxide.
    - genetic storage disorders are possible but poorly characterised in horses.
- Mycotoxins are common in an equine diet, especially in preserved forages:
  - mycotoxicosis has been associated with cereal feeds and fresh pasture plants (e.g. clovers).
  - aflatoxins and fumonisins are known to cause hepatotoxicity in horses.
  - commercially available services exist for determination of mycotoxin content of feedstuffs:

♦ incidental presence of hepatomycotoxins may be found in some forage samples fed to horses with no evidence of hepatic injury.
♦ large quantities of stored forage compared with the small size of samples submitted means considerable potential for sampling error.

Infectious hepatitis

- various infectious agents including bacteria, viruses and parasites may be implicated in liver disease.
- frequently seen as an outbreak in multiple horses on the same premises:
  ○ consider contagious causes alongside dietary/toxic factors.
- **Bacterial hepatitis**
  ○ sporadic cases of bacterial infection of the biliary tract with, or without, cholelithiasis (formed from calcium bilirubinate and calcium phosphate).
  ○ horses with bacterial cholangiohepatitis and cholelithiasis tend to present with a combination of pyrexia, jaundice and colic signs.
  ○ diagnosis with reasonable confidence on a clinical basis.
  ○ speculated that duodenal reflux via the bile duct leads to ascending biliary infection:
    ♦ frequent isolation of Enterobacteriaceae from liver biopsy specimens in cholangiohepatitis cases.
  ○ Tyzzer's disease (*Clostridium piliforme*) is a cause of sudden death in foals:
    ♦ terminal clinical signs may include depression, jaundice, diarrhoea and neurological signs.
- **Viral hepatitis**
  ○ *Flaviridae* may cause liver disease and include:
    ♦ Theiler's disease-associated virus.
    ♦ hepacivirus known as non-primate hepacivirus.
  ○ equine parvovirus and kirkovirus have come under suspicion but their clinical relevance remains unclear and under investigation.

- **Nematodes**
  ○ hepatic migration of larval stages of nematodes such as:
    ♦ *Parascaris equorum*, *P. univalens*, *Strongylus equinus* and *S. edentates*.
    ♦ may be associated with multifocal liver injury, ultrasonographic changes and increased biochemical markers of liver disease.
  ○ tends to affect younger horses.
  ○ rarely causes enough liver injury to result in clinical signs although acute haemorrhagic necrosis is reported.
- **Trematodes**
  ○ *Fasciola hepatica* (liver fluke) infection is an occasional cause of hepatitis in some specific geographical areas:
    ♦ general ill-thrift is reported in some cases.
    ♦ pathogenicity is minor in many cases with clinical impact mild to negligible.
    ♦ serum biochemical markers may remain within normal limits.
  ○ obstruction of bile ducts and associated chronic inflammation are important contributors to their clinical effects.
  ○ donkeys may be predisposed.
  ○ wet ground and a history of infection in cattle and sheep locally increases the index of suspicion.
  ○ faecal examination for *Fasciola* eggs or coproantigen:
    ♦ highly insensitive diagnostic methods.
    ♦ magnified by non-patency of infection in many cases.
  ○ serological tests described but vary in their diagnostic usefulness.
- **Cestodes**
  ○ Hepatic cystic echinococcosis (CE):
    ♦ infection of the liver with hydatid cysts.
    ♦ intermediate stage of the tapeworm *Echinococcus granulosus equinus*.
  ○ usually suspected based on characteristic ultrasonographic images:
    ♦ approximately spherical 20–70 mm diameter anechoic intrahepatic cysts with or without mobile echogenic sediment within (Fig. 1.17).

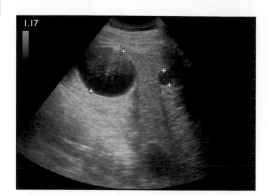

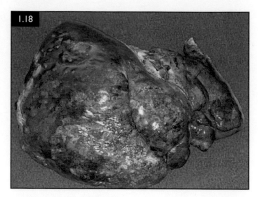

FIG. 1.17 A typical large cyst and a less typical smaller anechoic cyst in the liver of a mature pony affected by cystic echinococcosis. The deeper, less distinct anechoic structure is the hepatic portal vein.

FIG. 1.18 Hepatocellular carcinoma in a 6-year-old Thoroughbred mare.

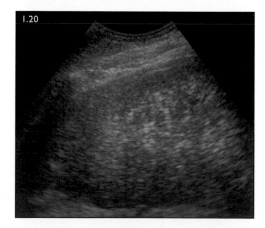

- single or multiple cysts may be present in the same horse:
  - usually without clinical consequence.
  - large burdens of multiple cysts have the potential to affect serum hepatic enzymes and/or liver function.
- domesticated hunting dogs that are fed on raw horse offal (e.g. fox hounds) are the main definitive host for the parasite.
- disease is international, but many countries appear free.
- most affected horses remain subclinically infected for life and tolerate the cysts well (occupy small volume of the liver).

FIG. 1.19 Focal mass of different echogenicity from normal liver in a case of biliary carcinoma.

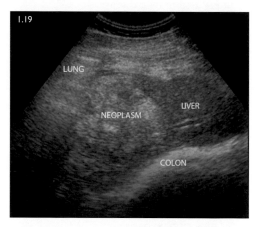

Neoplastic liver disease

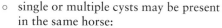

- hepatic neoplasia rarely encountered in the horse:
  - occasional site of both primary and metastatic neoplasms.
- most common primary tumours:
  - biliary carcinoma.
  - lymphoma.
  - haemangiosarcoma.
  - hepatocellular carcinoma (Fig. 1.18).
  - hepatoblastoma.
- main causes of metastatic neoplastic liver disease:
  - squamous cell carcinoma.
  - melanoma.

FIG. 1.20 Diffuse coarsely echogenic tissue within the liver of a horse with hepatic lymphoma.

- ultrasonographic changes often clearly abnormal in hepatic neoplasia (Figs. 1.19, 1.20):

- sometimes subtle in comparison to the gross post-mortem appearance.
- minor changes in regularity or diffuse echogenicity are suspicious.
- biopsies not specifically targeted by ultrasound may frequently miss neoplastic tissue.
- clinical signs include:
  - general signs of malaise and weight loss.
  - occasional cases of somnolence due to hypoglycaemia.
- absolute erythrocytosis recognised as a paraneoplastic syndrome in hepatoblastoma and hepatocellular carcinoma cases.

TREATMENT AND MANAGEMENT OF LIVER DISEASE

- **most crucial element when dealing with hepatic disease and/or insufficiency is to address any specific underlying cause of hepatic insult if it can be identified.**
- additional considerations include:
  - dietary management and the use of antimicrobials.
  - specific treatment of HE and hepatic fibrosis.

Dietary management

- horses with liver disease, in the absence of hepatic insufficiency, are unlikely to benefit significantly from dietary changes.
- where the liver is failing to perform its many nutritional and metabolic functions, there may be some benefit from supportive nutritional strategies.
- hepatic insufficiency associated with weight loss (protein-calorie malnutrition):
  - poor appetite.
  - insulin resistance.
- restriction of dietary protein promotes catabolism of endogenous proteins and should be avoided:
  - crude protein intake (1.2–1.5 g/kg of body weight [BWT] daily) is optimal in cases with hepatic insufficiency.
- when stabled and provided with forage and other feeds, a digestible energy (DE) intake of approximately 150–170 kJ/kg (35–40 kcal/kg) BWT should be provided:
  - adjusted up or down according to body condition.
- grazing should be encouraged:
  - accurate measurement of food intake is impossible.
  - may be problematic when photosensitisation is a concern.
  - *ad libitum* grass hay or haylage should be provided.

- frequent supply of dietary non-structural carbohydrates is beneficial in hepatic failure:
  - provide up to 0.5–1.0 g/kg BWT starch per meal.
  - cereal-based mixes fed every 6–8 hours.
  - whole or rolled oats (typically 50% starch) are preferred.
  - cooked or micronised maize (typically 70% starch) may be used.
  - commercial compounded concentrate mixes and 'sweet feeds' may also be used:
    - 15–25% starch, 8–10 MJ/kg (1.9–2.4 Mcal/kg) DE.
    - 8–10% crude protein.
- when frequent meals are not practical:
  - cereals mixed with forage (large buckets) may promote a constant slow intake of starch.
- sugar beet pulp is useful to provide good-quality fibre, with a small amount of cereal or molasses to provide starch or sugars.
- addition of vegetable oil (0.1–0.5 ml/kg BWT) increases ration energy density:
  - little evidence of serious fat malabsorption in most hepatic insufficiency cases.
- no basis for B vitamin supplementation in dietary management unless appetite and intake are poor.
- **care should be taken with supplements containing iron, which may promote further damage from haemosiderin accumulation in the liver.**
- supplementation of the hepatic-stored fat-soluble vitamins at 60 IU/kg vitamin A, 13 IU/kg vitamin D, 2 IU/kg vitamin E and 0.2 mg/kg vitamin K.

- zinc supplementation (e.g. 0.5 mg/kg) may be beneficial.
- nutritional supplements are popular for horses suffering from liver disease:
  ○ s-adenyl methionine (SAMe).
  ○ flavinolignans derived from milk thistle (*Silybum marianum*):
    ♦ collectively known as silymarin.
  ○ claimed to act as antioxidants and to protect against the effects of hepatotoxins.
  ○ evidence basis very weak, especially at low doses administered to horses.

Treatment of hepatic encephalopathy

- sedation is usually required in cases with moderate to severe signs of HE:
  ○ reduced doses of sedatives and tranquillisers:
    ♦ xylazine 0.5 mg/kg intravenous (i/v); detomidine 10 μg/kg i/v; acepromazine 25 μg/kg i/v.
- diazepam or midazolam avoided due to implication of benzodiazepine-like substances in the pathophysiology of HE.
- acute management of severe HE is best approached with i/v boluses of hypertonic (7.2%) saline (6 ml/kg) to reduce cerebral oedema.
- oral administration of lactulose (0.3 ml/kg q4–12 h):
  ○ lowers colon pH, thus decreasing absorption of ammonium ions.
  ○ increases incorporation of ammonia into bacteria.
  ○ inhibits luminal urease and proteases.
- oral antimicrobials including:
  ○ metronidazole (15–25 mg/kg q12 h) or neomycin (15 mg/kg q6 h).
  ○ used as a means of reducing intestinal bacterial ammonia genesis.
- other products used in humans but without any current evidence of safety or efficacy in horses include:
  ○ L-ornithine-L-aspartate (LOLA) to promote peripheral ammonia detoxification.
  ○ sodium benzoate and zinc supplementation:
    ♦ purported benefits on ammonia detoxification.

- ○ dopamine receptor agonist – bromocriptine.
  ○ benzodiazepine receptor antagonist – flumazenil.
  ○ glutamate receptor antagonists.
  ○ NSAIDs.
  ○ antihistamines.

Antifibrotic therapy

- hepatic fibrosis is common:
  ○ single histopathological feature with strongest association with a poor outcome.
- many proposed antifibrotic therapies exist but lack an evidence basis in horses.
- anti-inflammatory drugs are suggested to moderate the fibroplastic process and include:
  ○ prednisolone (1–2 mg/kg q24 h).
  ○ dexamethasone (0.05 mg/kg alternate days to 0.1 mg/kg q24 h).
  ○ pentoxifylline (10 mg/kg q12 h).
  ○ azathioprine (3 mg/kg q24 h).
- angiotensin-converting enzyme inhibitors may disrupt hepatic fibrosis and promote hepatic regeneration, e.g. benazepril (0.25–0.5 mg/kg p/o q24 h).
- Colchicine (0.03 mg/kg p/o) is a pyrrolizidine alkaloid:
  ○ used for many years as a potential hepatic antifibrotic drug in horses.
- antioxidants are also suggested to help control the fibrotic process and vitamin E (2–5 IU/kg daily) may be useful.
- phlebotomy to mobilise hepatic iron stores in cases of excessive haemosiderosis:
  ○ approximately 1.5 litres of blood per 100 kg BWT safely removed every 1–2 weeks.
  ○ haematology monitored and repeat biopsy to evaluate impact on hepatic iron.

Antimicrobials

- rarely indicated based on definitive diagnosis by biopsy.
- 3 main reasons for using antimicrobials in liver disease:
  ○ bacterial hepatitis (usually cholangiohepatitis).
  ○ hepatic abscesses.
  ○ reducing colonic ammonia genesis in HE cases (see above).

- bacterial cholangiohepatitis and hepatic abscesses require long-term treatment for several weeks.
- antibacterials suitable for enteral administration are preferred including:
  - potentiated sulphonamides (30 mg/kg combined product p/o q12–24 h),
  - enrofloxacin (5–7.5 mg/kg p/o q24 h),
  - doxycycline (10 mg/kg p/o q12–24 h),
  - metronidazole (20–25 mg/kg p/o q12 h).
- foals with Tyzzer's disease should be treated aggressively with sodium benzyl penicillin (50,000 IU/kg i/v q6 h) or oxytetracycline (10 mg/kg i/v q12 h).

Other drugs

- Fascioliasis may be treated with triclabendazole (12–15 mg/kg p/o):
  - possible activity against adults and larvae.
  - resistance may be common.
- Closantel – drug toxic to adult *Fasciola:*
  - used as an alternative agent for fascioliasis in donkeys.
  - safety has not been established in horses.
- hydatid cysts which are numerous and contributing to disease may be treated with:
  - albendazole (10 mg/kg p/o q24 h for 2–3 months).
  - cyst aspiration and injection of sterile saturated saline into the cyst cavity.
  - praziquantel has also been advocated in human medicine at a dose of 25–50 mg/kg once or twice weekly alone or in addition to albendazole.

Endocrine System

Diagnostic approach

History

- take a thorough history:
  - breed ○ age ○ primary complaint.
  - similar disorder in related animals.
  - activity and purpose of animal.
  - chronicity and progression of the disorder.
- diet:
  - type of feed ○ quantity ingested.
- water access and consumption.
- presence or history of other disorders:
  - sepsis ○ recurrent infections.
  - chronic lameness.
  - delayed wound healing.
- concurrent treatments:
  - protein-bound drugs such as NSAIDs can alter hormone concentrations.
- seasonality of the disorder.
- frequency and appearance of urination.

Physical examination

- complete physical examination is essential:
  - BWT ○ body condition score (BCS).
  - Cresty neck score.
  - fat distribution is important:
    - feel the fat deposits.
  - hair coat:
    - length ◆ condition.
    - areas of alopecia.
  - mental alertness and activity level.
- lameness examination, including hoof testers.
- foal:
  - examine musculoskeletal development.

Haematology

- complete blood count (CBC):
  - rule out infectious diseases.

- usually non-specific changes.
- serum biochemistry:
  - rule out an organ system disease.
  - include triglycerides, cholesterol, and electrolytes.

Urinalysis

- standard analysis including urine specific gravity.
- urinary fractional excretion (UFE):
  - useful in secondary hyperparathyroidism or adrenal insufficiency.
- urinary concentrations of specific neurotransmitters or hormones rarely performed.

Hormone analysis

- most common diagnostic tool.
- dynamic testing showing response of hormones to stimulation is more useful than baseline serum concentrations.
- many sources of variability:
  - individual and breed variation:
    - horse *vs* pony.
  - effect of diet:
    - fasted *vs*. grain *vs* hay.
  - geographical variations:
    - temperate *vs* tropical climates.
  - circannual rhythm:
    - autumn *vs* spring.
  - circadian variation:
    - morning *vs* afternoon.
  - hormone stability:
    - ACTH *vs* insulin.
  - assay used:
    - variation between laboratories.
- challenging interpretation:
  - standardise protocols to compare cases and detect improvement/ worsening.

DOI: 10.1201/9781003453666-2

DISEASES OF THE THYROID GLAND

Thyroid diseases in the adult horse

Definition/overview

- thyroid hormones:
  - thyroid gland traps iodide and uses it to produce/store primarily Thyroxin (T_4).
  - control of hormone release involves hypothalamus (thyrotropin-releasing hormone [TRH]), pituitary (thyroid-stimulating hormone [TSH]) and thyroid.
  - only free hormones (non-protein bound) are metabolically active.
  - Thyroxin (T_4) is primarily released:
    - ◆ de-iodinated to Triiodothyronine (T_3) in blood and tissues (more active).
  - involved in:
    - ◆ resting metabolic rate.
    - ◆ energy metabolism.
    - ◆ organ growth and maturation.
- hypothyroidism and hyperthyroidism are described in the horse:
  - true thyroid gland dysfunction is extremely rare.

Aetiology/pathophysiology

- true hypothyroidism is not well documented in the horse:
  - may result from thyroid tumours.
  - most of these are benign non-functional adenomas:
    - ◆ hormone levels within normal limits.
- **Primary hypothyroidism**
  - interference with iodine uptake by thyroid gland.
  - interference with thyroid hormone synthesis or release:
    - ◆ iodine deficiency.
    - ◆ selenium deficiency.
    - ◆ ingestion of goitrogens:
      - – sulphurated organic compounds, thiocyanates and isothiocyanates.
    - ◆ excessive maternal iodine ingestion during gestation can cause disease in neonatal foals.

- **Secondary hypothyroidism**
  - impaired TSH secretion.
  - usually with phenylbutazone or trimethoprim sulfadiazine administration.
- **Tertiary hypothyroidism**
  - impaired TRH secretion.
  - not described in horses.

Clinical presentation

- not well defined and historically confused with Equine Metabolic Syndrome.
- confirmed clinical signs include:
  - lethargy   ○ work intolerance.
  - alterations in hair coat.
  - decreased basal heart rate.
  - low cardiac output.
  - low respiratory rate and rectal temperature.
- increased triglycerides, cholesterol and very-low-density lipoproteins (VLDLs).
- all signs may be mild and not always outside normal values.

Differential diagnosis for goitre

- benign thyroid tumour (most commonly the case).
- other causes of swelling in the upper neck (enlarged salivary gland or lymph node).

Diagnosis

- 4 weeks without drug administration including thyroid supplements.
- resting T_4 and T_3 concentrations may not reflect actual status.
- thyroid stimulation tests are essential (Table 2.1).

Management

- hormone supplementation (Table 2.2).
- monitor serum thyroid hormone concentrations.

Prognosis

- good for hypothyroidism if properly diagnosed and treated.

TABLE 2.1 Protocol for thyrotropin-releasing hormone stimulation testing	
1	Baseline serum sample.
2	Administer 1 mg TRH/horse i/v (0.5 mg to ponies and foals).
3	Additional serum samples at 1, 2 and 4 hours, following administration.
4	Measure free T_3, total T_3, free T_4, total T_4 and TSH (if available).
5	TSH should peak after 1 hour and should increase to 2.5–3 times the baseline level.
6	Free and total T_3 should double 2 hours following TRH administration.
7	Free and total T_4 should increase by at least 1.7 times the baseline levels at 4–6 hours.
8	Time frame for increases in thyroid hormones in response to TRH is similar in foals to that in adults, but the amount of increase is less. • foals less than 3 days of age, total and free T_4 should increase approximately 10% 4 hours after administration of 0.5 mg TRH. • free T_4 after equilibrium dialysis should increase by approximately 30%. • total and free T3 should increase by 40% and 60%, respectively, 2 hours after TRH administration.

TABLE 2.2 Recommended doses of thyroxine and triiodothyronine for adult and neonatal horses
Adults • L-thyroxine: 20 µg/kg p/o q24 h (approximately 10 mg/adult horse q24 h) • Triiodothyronine: 1 mg/horse p/o q24 h
Foals • L-thyroxine: 20–50 µg/kg p/o q24 h. Monitor blood T_4 to avoid overdosing • Triiodothyronine: 1 µg/kg p/o q24 h

Thyroid diseases in the foal

Definition/overview

• foals exposed to an excess or deficiency of iodine *in utero*.

Aetiology/pathophysiology

• higher serum concentrations of thyroid hormones in neonates:
  ○ decreases gradually after birth.
  ○ adult reference range:
    ♦ free fractions within a few weeks.
    ♦ total fractions within a month.
• selenium deficiency and iodine-trapping plants (mustard) potential causes of goitre.
• excessive iodine ingestion by the mare whilst pregnant:
  ○ goitre and hypothyroidism in neonatal foals.

Clinical presentation may include

• goitre • incoordination • hypothermia.
• poor suckle and righting reflexes.
• syndrome described in foals in Western Canada and USA:
  ○ congenital musculoskeletal abnormalities including:
    ♦ mandibular prognathia (Fig. 2.1).
    ♦ flexural deformities of forelimbs and ruptured digital extensor tendons.
    ♦ incomplete ossification of carpal and tarsal bones (Figs. 2.2, 2.3).

Differential diagnosis

• weak foals, including dysmaturity and sepsis.

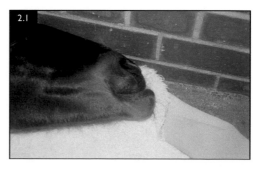

FIG. 2.1 Photograph of a foal with congenital hypothyroidism. Note the mandibular prognathia.

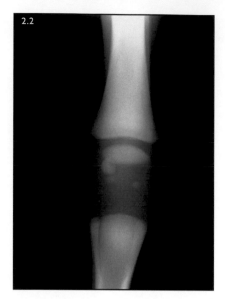

FIG. 2.2 Dorsopalmar radiograph of incomplete ossification of cuboidal bones of the carpus in a foal with congenital hypothyroidism.

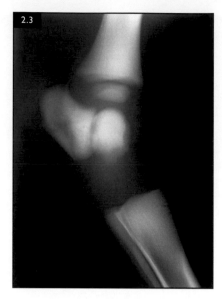

FIG. 2.3 Lateral radiograph of incomplete ossification of cuboidal bones of the tarsus in a foal with congenital hypothyroidism.

Diagnosis

- clinical signs of goitre.
- measurement of serum concentrations of thyroid hormones:
  - T_3 values are more consistent than T_4 in foals.
- TRH stimulation test.

Management

- correct iodine concentrations in diets fed to pregnant mares.
- thyroid hormone supplementation (see Table 2.2).
- restricted exercise in foals with incomplete cuboidal ossification (see Book 2).
- supportive care in recumbent foals.

Prognosis

- guarded for foals with goitre but better if survival for first week of life.
- foals with musculoskeletal deformities vary depending on the severity of the problems.

Hyperthyroidism

Overview

- extremely rare:
  - thyroid hormone-producing tumours.
  - deliberate or accidental overdose of oral thyroxine administration.
- clinical signs include:
  - weight loss ○ tachycardia/tachypnoea.
  - hyperactivity.
  - ravenous appetite.
- treatment by thyroidectomy or hemithyroidectomy, followed by replacement hormone therapy is successful in treating neoplastic cases:
  - correct dosing of hormone supplementation is curative for iatrogenic cases.

Diseases of the parathyroid gland

Definition/overview

- hyperparathyroidism is more frequent than hypoparathyroidism:
  - primary hyperparathyroidism due to parathyroid gland adenoma (Fig. 2.4).
- secondary hyperparathyroidism more common and nutritional in most cases:
  - bone demineralisation caused by imbalance of calcium/phosphorous in diet.
  - young, growing horses and lactating mares are more at risk.

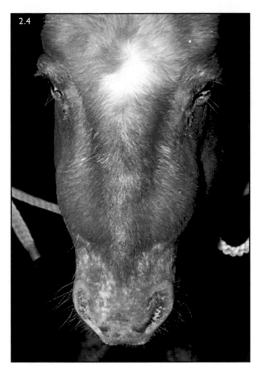

FIG. 2.4 Advanced fibrous osteodystrophy (big head) in a horse with primary hyperparathyroidism due to a PTH-secreting tumour.

TABLE 2.3 Diets associated with increased risk of nutritional secondary hyperparathyroidism

- Grain with high phosphorous content and poor to average hay (i.e. oats [Ca 0.07%, P 0.37%] and grass hay [Ca 0.3%, P 0.3%]).

- Increased grain to roughage ratio

- Supplementation with rice or wheat bran (i.e. rice bran [Ca 0.04%, P 1.8%] and wheat bran [Ca 0.12%, P 1.43%])

- Pasture or hay with oxalate:calcium ratios >0.5% (i.e. *Setaria*, Argentine or Dallas grass, and buffalo grass)

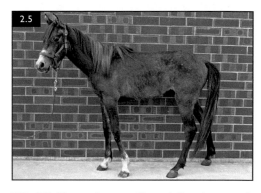

FIG. 2.5 Young horse with nutritional secondary hyperparathyroidism. Note the thin, unthrifty appearance and the enlarged head.

Aetiology/pathophysiology

- Nutritional secondary hyperparathyroidism:
    - improper diet, deficient in calcium or excessive in phosphorous.
    - oxalate-containing plants decreasing calcium absorption.
    - specific imbalanced diets (Table 2.3).
    - low serum calcium or high phosphorous increases PTH and active vitamin D.
    - increased calcium and phosphorous resorption from the gastrointestinal tract and bones.
    - increased renal tubular excretion of phosphate.
    - maintains normal serum calcium (and phosphorus) concentration at expense of bone.
- Primary hyperparathyroidism:
    - parathyroid adenoma with excessive PTH secretion.
- Primary hypoparathyroidism:
    - idiopathic condition resulting in life-threatening hypocalcaemia.

Clinical presentation

- Nutritional secondary hyperparathyroidism:
    - unthrifty appearance (Fig. 2.5).
    - bone demineralisation leads to a stiff gait, lameness and painful joints:
        - pathological fractures.
        - loose teeth and painful mastication.
    - firm enlargements of the facial bones dorsocaudal to the facial crest.
    - thickening of the mandibles and nasal bones, with possible nasal obstruction.
- Primary hyperparathyroidism and hypercalcaemia:
    - vague clinical signs such as poor appetite and shifting leg lameness.
- Primary hypoparathyroidism:
    - see hypocalcaemia (below).

TABLE 2.4 Formula for calculating urinary fractional excretion of electrolytes

- Clearance ration (X) = $\dfrac{[x] \text{ urine } [\text{creatinine}] \text{ serum } " \times 100}{[x] \text{ serum} \times [\text{creatinine}] \text{ urine}}$

- Normal % clearance: Na: 0.02–1.0; K: 15–65; PO_3: 0.0–0.5

- PO_3 excretion >4%, nutritional hyperparathyroidism is present

Differential diagnosis for hypercalcaemia

- chronic kidney disease.
- primary hyperparathyroidism.
- nutritional secondary hyperparathyroidism.
- other causes of unthriftiness and lameness in young animals.
- vitamin D toxicoses.
- pseudohyperparathyroidism of malignancy.

Diagnosis

- submit blood for a PTH panel:
  - measures PTH, ionised calcium and PTH-related peptide.
  - increased with pseudohyperparathyroidism of malignancy.

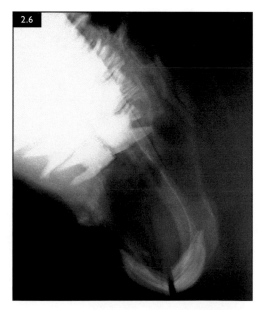

FIG. 2.6 Lateral radiograph of the skull of the horse in Fig. 2.5. Demineralisation of the skull is apparent by the marked difference in opacity between the bones of the skull and the cheek teeth.

- review complete dietary history:
  - calcium and phosphorous ratio.
  - serum calcium and phosphorus levels will be normal.
- radiography of bones:
  - signs when 30% demineralised:
    - skull more sensitive than limbs (Fig. 2.6).
  - decreased density of the laminae durae dentes first then facial bones.
- increased fractional excretion of phosphorous in the urine (Table 2.4):
  - normal fractional excretion of phosphorous is 0.0–0.5%.
  - 0.5% suggestive   >4% diagnostic.
  - perform test before any switch in diet.

Management

- oral calcium supplementation (Table 2.5):
  - commercial mineral mix or limestone.
  - mild case, normal balanced diet with Ca/P at 2/1.
  - severe case, supplementation of diet with Ca/P at 4/1.
- restriction of exercise to limit the risk of pathological fractures.
- judicious use of anti-inflammatory drugs if painful or severe lameness:
  - excessive use can increase the risk of fractures.

Prognosis

- depends on severity of disease at diagnosis:
  - good for mild cases with complete recovery.
  - fair for moderately affected with soundness, but facial enlargement remains.
  - guarded for severely affected horses with lameness, upper respiratory obstruction, dysmastication or pathological fractures.

TABLE 2.5 Recommended daily intake of calcium for normal horses and foals				
	% IN DIET		**DAILY REQUIREMENT (g)**	
	Ca	**P**	**Ca**	**P**
Foal < 6 months	0.80	0.55	33	20
Weanling	0.60	0.45	34	25
Yearling	0.50	0.35	31	22
Two-year-old	0.40	0.30	25	17
Mare, late pregnancy	0.45	0.30	34	23
Mare, lactation	0.45	0.30	50	34
Adult, maintenance	0.30	0.20	23	14

From Schryver HF, Hintz HF (1987) Minerals. In: *Current Therapy in Equine Medicine 2*. (ed. NE Robinson) WB Saunders, Philadelphia, p. 396.

Hypocalcaemia

Definition/overview
- decreased blood calcium concentration caused by, or associated with, a variety of conditions (Table 2.6):
  - primary hypoparathyroidism is an uncommon cause.

Aetiology/pathophysiology
- sudden changes in ionised calcium:
  - decrease in calcium intake
  - increased calcium demand.
  - lactation.
  - increase in calcium loss in faeces, urine or sweat.
  - sequestration of calcium.
- extracellular calcium ions act as a sodium channel antagonist by decreasing sodium permeability and increasing the depolarisation threshold.
- decrease in calcium concentration leads to hyperactivity of the sodium channels.
- calcium and magnesium homeostasis are linked:
  - not unusual for hypocalcaemia cases to have low magnesium concentrations.

Clinical presentation
- varies depending on severity of hypocalcaemia.
- signs associated with ionised calcium level not total calcium.
- increased muscle tone and activity.

- **mild signs:**
  - colic due to decreased gut motility and gas distension.
  - synchronous diaphragmatic flutter.
  - hyperexcitability.
  - tachypnoea.
  - tachycardia, with or without arrhythmia.
- **moderate signs:**
  - tetany ○ incoordination ○ gait stiffness.
  - abnormal facial expressions.
  - elevation of the tail.
  - muscle fasciculations particularly in neck, trunk and upper limb muscles.
  - laryngospasm   ○ profuse sweating.
  - dehydration.
  - pyrexia   ○ dysphagia.
- **severe signs:**
  - recumbency   ○ stupor.
  - seizures   ○ death.

Differential diagnosis
- depends on severity of clinical signs:
  - colic   ○ tetanus.
  - hyperkaliaemic periodic paralysis (HYPP).
  - myositis.
  - equine motor neuron disease.
  - other neurological diseases.
  - exhausted horse syndrome.

Diagnosis
- clinical signs, history of predisposing conditions and response to treatment.

TABLE 2.6 Conditions associated with hypocalcaemia

Pregnancy/lactation
- Lactation in the mare ('eclampsia' or 'lactation tetany').
- Most commonly 10–86 days after parturition at the peak of lactation.
- Uncommonly late gestation or 1–2 days after weaning.
- Mares eating diets low in calcium, or performing physical work are at high risk.

Sweating (loss of fluid and electrolytes)
- Endurance events – exhausted horse syndrome.
- Prolonged transport, especially in heat and humidity.
- Hot, humid environments.

Alkalaemia
- Association with K and Cl loss in sweat.
- Hypokalaemia associated with anorexia.
- Hypochloraemia with severe gastric reflux.
- Respiratory alkalosis caused by hyperventilation.

Sepsis/endotoxaemia
- Gastrointestinal disease
- Metritis
- Pleuropneumonia
- Retained placenta
- Increased procalcitonin, perhaps an inflammatory cytokine

Primary hypoparathyroidism
Hypomagnesaemia
Acute kidney injury
Acute rhabdomyolysis
Urea poisoning
Hepatitis
Malabsorption syndromes
Cantharidin toxicosis
Rapid intravenous tetracycline administration
Corticosteroids
Idiopathic (miniature horses seem to be susceptible to this condition)

- serum ionised calcium levels are definitive and should be ≥10 mmol/l:
  - signs become obvious below 8 mmol/l and severe below 5 mmol/l.
  - calcium is protein bound in plasma:
    - ♦ level that is bound varies with the acid–base status:
      - – measure acid–base status.
      - – alkalosis decreases and acidosis increases ionised fraction.
  - plasma protein levels will affect the level of total calcium in the blood:
    - ♦ hypoalbuminemia decreases total calcium levels.
- only total serum calcium levels are available:
  - correct total serum calcium (mg/dl) = measured serum calcium (mg/dl) − serum albumin (g/dl) + 3.5.
- other abnormalities seen in association:
  - metabolic alkalosis.
  - hypo/hypermagnesaemia.
  - hypo/hyperphosphataemia.
  - dehydration.
  - other electrolytes (P, K, Cl, Na and Mg) should also be assessed.

Management
- daily requirements for normal adult: 40 mg/kg per day.
- dietary management – feed adequate, not excessive, amounts of calcium in the diet.
- pregnant, lactating mares or horses participating in endurance exercise need more.
- **mild cases or prevention:**
  - oral supplementation with limestone or alfalfa hay:
    - ♦ 20–60 g limestone orally over several hours.

◆ paste in the mouth or sprinkled on feed.
◆ do not over-supplement.
- **more severe cases:**
  ○ intravenous calcium gluconate +/– dextrose and electrolytes, calcium borogluconate or calcium chloride.
  ○ 500 ml of 23% calcium gluconate in 5 l of isotonic crystalloids over 1–2 hours:
    ◆ repeated as required.
  ○ monitor heart rate and rhythm:
    ◆ stop administration if bradycardia or arrhythmia.
    ◆ resolution of clinical signs in minutes to hours.
  ○ some cases may not resolve until they receive magnesium supplementation.
  ○ ionised calcium and magnesium levels monitored frequently to assess response to therapy.

Prevention

- analysis of the mineral content of feed and correction of Ca/P imbalances:
  ○ excess phosphates in the diet prevent absorption of calcium.
- predisposed horses and lactating mares should be protected from unnecessary stress or exertion.

Prognosis

- excellent if the cause is identified and properly managed.
- unrecognised, severe cases: poor; death can occur.
- primary hypoparathyroidism managed successfully by oral calcium supplement.

Pituitary pars intermedia dysfunction (PPID)

Definition/overview

- very common disease of older horses (>15 years old).
- adenoma or adenomatous hyperplasia of the pituitary gland.
- associated with laminitis and recurrent infections.

Aetiology/pathophysiology

- neurodegeneration of the inhibitory dopaminergic hypothalamic neurons.
- unknown cause but genetic predisposition in certain breeds may be a factor.
- results in adenomas or adenomatous hyperplasia of the pars intermedia melanotroph population:
  ○ overproduction of pro-opiomelanocortin (POMC) cleaved into:
    ◆ adrenocorticotropic hormone (ACTH).
    ◆ α-melanocyte-stimulating hormone (α-MSH).
    ◆ corticotropin-like intermediate peptide (CLIP).
    ◆ α-endorphins.
  ○ no increase in serum baseline cortisol level.

Clinical presentation

- long shaggy coat that does not shed normally in the spring (hypertrichosis) (Fig. 2.7).
- older horse (>15 years old) with lethargy and a pendulous abdomen.

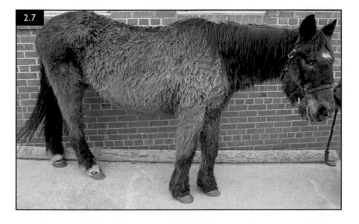

FIG. 2.7 Typical appearance of a horse with PPID. Note the long curly hair coat and poor muscling.

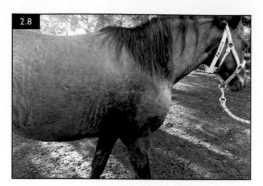

FIG. 2.8 Hyperhidrosis in a horse with PPID. Sweating is present despite a relatively normal haircoat.

TABLE 2.7 Protocol for the TRH stimulation test	
1	Collect baseline plasma sample for ACTH concentration.
2	Administer TRH (1 mg i/v, or 0.5 mg for a pony).
3	Collect a plasma sample for ACTH concentration at 10 and/or 30 minutes. Plasma ACTH concentration will be increased to >110 pg/ml (10 minutes) or >65 pg/ml (30 minutes) in horses with PPID.
4	Values need to be adapted based on location and season.

- less advanced cases maybe more subtle:
  - poor performance.
  - muscle atrophy especially along the topline.
  - hoof changes with or without laminitis.
  - predisposition to developing solar abscesses.
  - delayed coat shedding.
  - loss of weight.
  - recurrent laminitis.
  - recurrent infections (respiratory, alimentary, ocular, cutaneous).
  - non-healing skin wounds.
  - dental and sinus disease.
  - hyperhidrosis, secreting sweat that tends to be greasy (Fig. 2.8).
  - subfertility in mares.
  - polyuria and polydipsia.
  - rarely neurological disorders in advanced cases due to expansion of the mass:
    - circling, blindness and seizures.

Differential diagnosis

- appearance of PPID is unique.
- recurrent laminitis is seen in equine metabolic syndrome (EMS).
- hair coat abnormalities.

Diagnosis

- clinical picture.
- baseline ACTH concentration:
  - circannual variability.
  - higher concentrations in autumn.
  - relative stability   o keep sample at 4°C.

- analyse within 48 hours.
- contextual interpretation:
  - several diagnostic cut-off values and reference ranges published.
- TRH stimulation test (Table 2.7):
  - increased sensitivity.
  - same limitations as above regarding stability.
- advanced imaging techniques may visualise pituitary abnormalities:
  - CT and MRI.
  - rarely used as require general anaesthetic (GA)/heavy sedation and lack sensitivity in early cases.
- serum cortisol concentration is not useful.
- **No longer recommended:**
  - dexamethasone suppression test.
  - domperidone stimulation test.
  - **note that diagnostic test results, when performed on multiple occasions, vary within an individual horse and interpretation must be carried out in a clinical context.**

Management

- Pergolide mesylate is the recommended treatment:
  - type-2 dopamine agonist that restores dopaminergic inhibition.
  - 2 µg/kg (1 mg/horse) once a day orally.
  - clinical and hormonal improvement within 1 to 3 months.
  - no improvement:
    - increase gradually by 1 µg/kg to up to 14 µg/kg every 6 to 8 weeks.
  - anorexia reported at high doses.

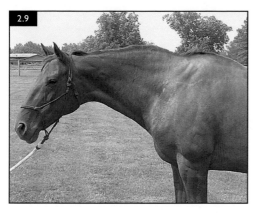
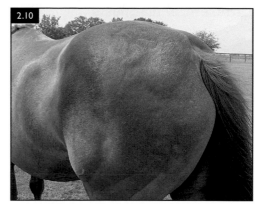

FIGS. 2.9, 2.10 A horse with equine metabolic syndrome displaying a typical cresty neck (2.9) and excessive fat deposition around the tail head area (2.10).

○ medications are not curative, and response only lasts if medication is given.
- Cabergoline is a type-2 dopamine agonist with the same method of action as pergolide:
  ○ available as a long-acting i/m injection given every 7–14 days in the USA.
  ○ anecdotal evidence of effectiveness at 0.005 mg/kg.
- investigate and address co-morbidities:
  ○ laminitis requires farriery every 4 weeks.
  ○ recurrent infections with antimicrobials.
  ○ clip the excess hair as needed.
  ○ provide regular dental care every 6 months.
  ○ careful attention to diet.
  ○ ensure adequate deworming and vaccination schedule.

Prognosis

- guarded:
  ○ better if diagnosed early prior to secondary complications.
- guarded to poor if severe laminitis and recurrent infections.

Equine metabolic syndrome (EMS)

Definition/overview

- collection of clinical signs and metabolic dysregulations:
  ○ insulin dysregulation (hyperinsulinaemia and/or insulin resistance).
  ○ fat deposits (generalised obesity or regional adiposity).
  ○ abnormal lipid metabolism (triglycerides, leptin, adiponectin).
  ○ additional features (hypertension and subfertility).
- results in a predisposition toward laminitis.
- major concern of the equine industry.
- EMS and PPID are two different, but not mutually exclusive, entities.

Aetiology/pathophysiology

- genetic predisposition:
  ○ ponies, Morgans, Paso Finos, Mustangs and Arabians.
  ○ all horses and ponies can develop EMS with an incorrect diet.
- high absorption of non-structural carbohydrates.
- insulin dysregulation:
  ○ tissue insulin resistance.
  ○ basal hyperinsulinaemia.
  ○ post-prandial hyperinsulinaemia.
- hyperinsulinaemia associated with laminitis but mechanism still unknown.

Clinical presentation

- thick, cresty neck (cresty neck score ≥ 3/5).
- fat accumulation (high body condition score ≥ 7/9 (Figs. 2.9, 2.10).
- chronic, low-grade, recurrent laminitis.

Differential diagnosis

- PPID:
  - two diseases can coexist in the same animal.
- type 2 diabetes mellitus if persistent hyperglycaemia.

Diagnosis

- Hyperinsulinaemia:
  - measuring baseline insulin has poor sensitivity.
  - dynamic tests are better and recommended:
    - predict the risk of laminitis.
    - oral glucose test (Table 2.8) and oral sugar test (Table 2.9).

TABLE 2.8 Protocol for the oral glucose test

1	Collect baseline serum sample for insulin concentration after at least 3 hours of no grain or concentrates
2	Administer glucose or dextrose powder (0.75 g/kg p/o) in chaff or by nasogastric tube
3	Collect a serum sample for insulin concentration 120 minutes later serum insulin concentration will be increased to <45 mIU/l in normal horses

TABLE 2.9 Protocol for the oral sugar test

1	Collect baseline serum sample for insulin concentration after at least 3 hours of no grain or concentrates
2	Administer corn syrup (0.15 ml/kg p/o or 0.45 ml/kg p/o)
3	Collect a serum sample for insulin concentration 60–90 minutes later serum insulin concentration will be increased to <45 mIU/l in normal horses using the low dose and <65 mIU/l using the high dose (more sensitive)

- peripheral tissue insulin resistance:
  - baseline glucose:
    - indirect measure of insulin resistance.
    - poorly specific and sensitive.
- two-step insulin response test (Table 2.10):
  - more sensitive but risk of hypoglycaemia.

TABLE 2.10 Protocol for 2-step insulin response test

1	Collect baseline blood sample for glucose concentration (no fast and access to hay)
2	Administer regular insulin (0.1 mIU/kg i/v, about 0.5ml for a 500-kg horse)
3	Collect a blood sample for glucose concentration 30 minutes later blood glucose concentration will be increased by >50% in normal horses and smaller or delayed decreases are observed in horses with peripheral tissue insulin resistance
4	Monitor for signs of hypoglycaemia for 30–60 minutes, and/or inject 50% dextrose i/v, or provide a high-carbohydrate meal after collection of the second sample

- combining the oral glucose/sugar test with the two-step insulin response test improves the detection of insulin dysregulation.

Management

- reduce absorption of non-structural carbohydrates:
  - feed 1.5–2% of body weight (dry matter) with hay low in non-structural carbohydrates:
    - <12% non-structural carbohydrate.
    - soaked for 30 minutes in hot water or 1 hour in cold water to reduce non-structural carbohydrate content.
- no pasture access initially and no grain or concentrates.
- mineral supplementation may be required.
- severe dietary restrictions can lead to hyperlipaemia:
  - restrictions below 1.5% of body weight (dry matter) are not recommended and should be monitored closely.
  - prolonged periods of fasting (over 6 hours) can lead to insulin resistance.
- increase exercise:
  - difficult in laminitic animals.
  - increases insulin sensitivity and weight loss.
- provide proper foot care with close co-operation with farrier:
  - radiographs are required for proper trimming.

- medical treatments:
  - Levothyroxine (0.1 mg/kg orally q24 h):
    - thyroid hormone supplement.
    - indirect effect leads to decrease in insulin levels:
      - increase metabolic rate, potentiating weight loss and fat metabolism.
    - used until target weight is reached and gradually reduced over a few weeks.
  - Metformin (15–30 mg/kg 2–4 times daily):
    - decreases post-prandial hyperinsulinaemia.
    - variable effects and becoming less popular.
  - sodium glucose co-transport inhibitors (SGLT2):
    - recently used with encouraging results.
    - inhibit re-uptake of glucose in the kidney and increasing excretion in urine.
    - leads to lower levels of insulin concentration in blood.
    - Velagliflozin   Canagliflozin.
    - Ertugliflozin (0.05 mg/kg orally q24 h).
    - reduces risk of laminitis.

Prognosis

- depends on the severity and frequency of laminitic episodes.
- euthanasia may be necessary in severe cases.
- requires a long-term plan with improvement over 6 to 12 months:
  - early detection is key in improving prognosis.

Hyperlipaemia and hyperlipidaemia

Definition/overview

- excessive fatty acids in the serum leading to a milky appearance.
- hyperlipaemia – serum triglycerides >6.5 mmol/l (500 mg/dl).
- hyperlipidaemia – milder increase with no visible serum changes.

- more common in ponies, donkeys and miniature horses.

Aetiology/pathophysiology

- negative energy balance results in activation of hormone-sensitive lipase.
- mobilisation of glycerol, free fatty acids (FFAs) and non-esterified fatty acids (NEFAs) from adipose tissue.
- converted in liver to glucose by gluconeogenesis or oxidised to acetyl coenzyme A (acetyl-CoA):
  - some re-esterified to triglycerides and very-low-density phospholipids.
  - accumulate in the hepatocytes leading to fatty liver.
- peripheral tissues cannot use the increased amounts of fatty acids or liver metabolites and therefore accumulate in serum.
- occasional primary disease in ponies, donkeys and miniature horses.
- more commonly, secondary to another disorder that creates a negative energy balance:
  - overweight individuals with tissue insulin resistance during periods of anorexia or feed restriction.
  - mares in late pregnancy or lactating.
  - anorexic animals with systemic diseases.
  - azotaemic animals at increased risk due to decreased tissue uptake of triglycerides.

Clinical presentation

- anorexia, depression and weakness.
- progresses to muscle fasciculations, ataxia, head-pressing and circling.
- recumbency, convulsions and coma.
- ventral oedema.
- diarrhoea is present either as an inciting cause or clinical sign.
- vague and insidious if caused by a systemic disease.

Differential diagnosis

- clinical signs are non-specific and caused by many diseases, so easily missed.

Diagnosis

- serum triglycerides >6.5 mmol/l or 500 mg/dl.
- increased liver enzyme activity.

- additional testing to identify possible inciting causes.

Management

- primary goals of treatment:
  - identify and treat inciting cause of negative energy balance or underlying disease.
  - improve the energy balance.
- mild cases:
  - offer palatable feeds.
  - administration of NSAIDs if pain or fever are present.
  - provision of higher-energy diet, force-feeding via nasogastric tube or dosing syringe.
  - glucose and electrolyte home-made solutions:
    - improperly balanced solutions may exacerbate metabolic acidosis or result in hyperglycaemia.
    - molasses, corn syrup, gruels made from pelleted feeds or commercially available enteral solutions.
    - start with smaller amount of feed (50%) and gradually increase.
    - multiple feeds per day.
    - usually enough to stimulate insulin secretion and inhibit hormone-sensitive lipase.

- moderate cases:
  - intravenous infusion of glucose (5 or 10%).
  - force-feeding by nasogastric tube.
  - balanced electrolyte intravenous fluid therapy:
    - maintenance and ongoing losses, especially azotaemic animals.
  - insulin if hyperglycaemic.
  - Heparin (40–250 USP units/kg s/c q12 h) enhances removal of lipids from blood.
- severe cases:
  - partial or total parenteral nutrition without the lipid component:
    - useful in GI tract disorders.
    - frequent monitoring required of blood glucose and triglycerides.

Prognosis

- worse in cases with severe underlying disorders which are difficult to treat.
- guarded to poor but improves when triglycerides return to the normal range within 3–10 days of treatment.
- prevention and close monitoring of animals at risk provide best results.

MISCELLANEOUS DISEASES OF THE ENDOCRINE SYSTEM

Diabetes mellitus

- Type 1 and type 2 diabetes mellitus both described in horses:
  - rare case reports of type 1.
  - few cases of type 2:
    - sustained hyperglycaemia, glycosuria, polyuria and polydipsia.
- rule out other problems:
  - Equine Metabolic Syndrome.
  - Pituitary Pars Intermedia Dysfunction.
  - renal disorders.

Adrenal insufficiency

Definition/overview

- adrenal hormones essential for maintenance of normal homeostatic mechanisms and response to stress:

  - glucocorticoids and mineralocorticoids.
- failure of the adrenal gland to respond despite ACTH secretion is primary insufficiency.
- relative adrenal insufficiency (RAI) or critical illness-related corticosteroid insufficiency (CIRCI):
  - basal cortisol secretion but failure to respond to stimulation in severe stressor situation such as sepsis.
  - both adult horses and neonatal foals.

Aetiology/pathophysiology

- poorly understood:
  - pathological process in the adrenal gland.
  - exhaustion of synthetic capacity.
  - decreased sensitivity of target tissues.

○ high ACTH concentrations in the face of normal or low cortisol concentrations and an inability of the adrenal gland to respond to stimuli.

Clinical presentation

- animal responding poorly to treatment or having more severe signs of a systemic inflammatory or infectious disease:
  ○ sepsis   ○ colitis.
  ○ pleuropneumonia   ○ colic.
- often vague signs including poor appetite, weakness, weight loss and polyuria/ polydipsia (PU/PD).

Differential diagnosis

- Foal:
  ○ sepsis ○ ruptured bladder ○ enteritis.
- Adult:
  ○ sepsis ○ internal abscess ○ poor diet.
  ○ parasitism.
  ○ colitis   ○ renal disease.

Diagnosis

- hyponatraemia and hypochloraemia with hyperkalaemia:
  ○ not always present if mineralocorticoids not involved.
- stimulation tests (Table 2.11):
  ○ determination of whether the adrenal glands can respond.

	TABLE 2.11 Protocol for ACTH stimulation testing
1	Take a control serum sample
2	Administer 0.1 µg/kg cosyntropin (Cortrosyn®) i/v
3	Take additional serum sample 30 minutes after
4	Submit samples for cortisol determination
5	30-minute sample should be double the baseline value of cortisol

Management

- corticosteroids rapidly reverse clinical signs:
  ○ acute cases dexamethasone 25 mg/450 kg/24h i/v.
  ○ hydrocortisone 1–4 mg/kg/24h i/v is useful in CIRCI cases.
  ○ prednisolone 0.25–1.0 mg/kg/24h p/o for long-term replacement in adults/ foals.
  ○ limited evidence for improved outcome.

Prognosis

- depends on the primary condition but can be good to fair.
- corticosteroid therapy gradually decreased and eventually stopped as normal adrenal function returns.

Adrenal tumours

Definition/overview

- uncommon, usually non-functional and incidental findings at necropsy.
- adrenal cortex tumours include adenoma and carcinoma.
- adrenal medulla tumours are commonly phaeochromocytoma.
- pathology caused by over-secretion of catecholamines.

Clinical presentation

- adrenal cortex tumours associated with colic, weight loss, limb oedema and seizures.
- phaeochromocytoma signs include:
  ○ none.
  ○ hyperhidrosis ○ muscle fasciculations.
  ○ tachycardia and tachypnoea.
  ○ polyuria and polydipsia   ○ colic.

Differential diagnosis

- systemic inflammatory response syndrome.
- colic.
- electrolyte abnormalities   • myositis.

Diagnosis

- phaeochromocytoma is via clinical signs and persistent hyperglycaemia:
  - confirmed by increased blood and/or urinary adrenaline (epinephrine).
- adrenal tumours may be palpable transrectally if large and on left side.
- abdominal ultrasound may reveal a mass near the kidney and direct a biopsy.
- most are diagnosed at necropsy.

Management

- surgical removal if unilateral, small and early diagnosis:
  - rarely performed as tumour too large when clinical signs are present.

Prognosis

- poor if clinical signs are present.

Urinary System

APPROACH TO A URINARY TRACT DISEASE

General physical examination

- prior to examination a thorough history should be taken:
  - duration and type of clinical signs.
  - medication(s) given and response to treatment.
  - diet, including quantity and quality.
  - reproductive status.
  - number of animals on the premises that are affected.
- full physical examination with specific attention to urinary tract:
  - clinical signs of urinary tract diseases are non-specific.
  - many urinary tract diseases resemble acute and/or chronic lower GI diseases.
  - male patients:
    - urethral orifice and urethra on ventral aspect of the penis should be palpated.
  - female patients:
    - vaginal opening and perineum can be examined.
  - foals:
    - umbilicus and surrounding abdomen should be assessed.
  - all animals:
    - assess hydration status and observe urination (Table 3.1).
- palpation per rectum performed in all animals of adequate size to assess:
  - proximal urethra (located cranial to the pelvic brim).
  - bladder, ureters and the caudal pole of the left kidney.
- water consumption and urine output should be assessed ideally over a 24-hour period.
  - can vary with age and may be influenced by climate, diet and level of exercise.
  - normal renal function produces 5–20 litres of urine daily (15–30 ml/kg/day):
    - whilst consuming 15–60 ml/kg/day.
    - increased water loss should result in increased water intake:

TABLE 3.1 Terminology used in urinary tract disease

Stranguria	Slow and painful urination
Pollakiuria	Abnormally frequent passage of urine
Dysuria	Painful or difficult urination
Oliguria	Reduced daily urine output
Polyuria	Increased daily urine output
Anuria	Complete cessation of urine output
Urinary incontinence	Uncontrolled (involuntary) leakage of urine
Azotaemia	Presence of nitrogenous waste products in the blood
Isosthenuria	Urine has the same SG/osmolality as plasma (1.008 and 1.012; 280–350 mOsm/kg)
Hyposthenuria	Urine SG/osmolality less than plasma (<1.008; <280 mOsm/kg)

– pathological (i.e. diarrhoea, haemorrhage, polyuria).
– physiological (i.e. sweating).

Diagnostic tests

Urinalysis

- essential to assess physical, chemical and microscopic characteristics of urine sample.
- results may be influenced by the method of urine collection:
  o voided, mid-stream urine samples can be readily obtained:
    ♦ contamination from distal urethra and genital tract makes interpretation of results more difficult.
  o early stream of voided urine can often distinguish:
    ♦ urethral and genital tract diseases from those within the bladder or proximal urinary system.
- urethral catheterisation is easy to perform in most horses (Fig. 3.1):
  o performed when voided urine cannot be collected or samples for bacterial culture are required (Fig. 3.2).
  o alpha-2 adrenergic agonists (i.e. xylazine, detomidine) and exogenous corticosteroid administration prior to collection cause diuresis and glucosuria.
  o trauma to the urinary tract is inevitable with catheterisation:
    ♦ often causes mild increase in urine protein levels, red blood cells (RBCs) and transitional epithelial cells.

Physical and chemical properties of urine

- equine urine should be pale yellow to deep tan (Fig. 3.3):
  o exposure to air usually causes the colour to darken.
  o large amounts of calcium carbonate crystals and mucus may be present in normal urine, resulting in a highly turbid sample.
- discolouration of urine can be caused by pigmenturia or haematuria (red to brown urine) and is abnormal:

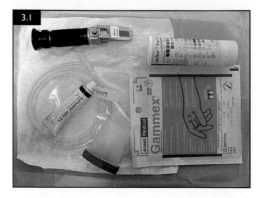

FIG. 3.1 Sterile gloves, sterile urinary catheter, sterile ointment or jelly, sterile cup, refractometer and urine test strips are basic equipment required for aseptic urine collection.

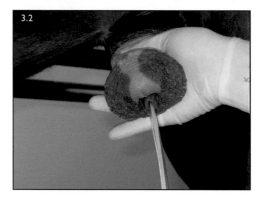

FIG. 3.2 Catheterisation of the urinary bladder in a male.

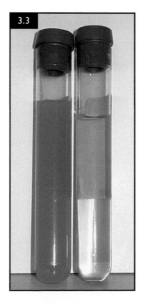

FIG. 3.3 Normal equine urine can be quite variable, ranging from clear to turbid.

- pigmenturia – presence of haemoglobin or myoglobin.
- haematuria – presence of intact RBCs.
- pigmenturia/haematuria in urine yields a positive result on reagent strips.
- differentiation of pigmenturia requires:
  - evaluation of the urine sediment for erythrocytes.
  - ammonium sulphate precipitation test to detect myoglobin.
  - protein electrophoresis differentiates haemoglobin from myoglobin.
  - serum samples should be analysed for concurrent haemolysis.
- several drugs may cause discolouration of urine (Table 3.2).
- discolouration of urine will be more evident in dehydrated horses.

TABLE 3.2 Drugs potentially causing discoloured urine

Acepromazine	pink to red-brown
Clofazimine	red to orange to brown
Furazolidone	dark yellow to brown
Metronidazole	red to brown
Nitazoxanide	bright orange to dark yellow
Phenazopyridine	red to orange
Rifampin	red to orange

- urine-specific gravity (SG) should always be measured using a refractometer:
  - horse with normal renal function and normal water intake should concentrate urine to SG between 1.018 and 1.025 (600–900 mOsm/kg).
  - dehydrated horse, SG may increase to/exceed 1.045 (~1500 mOsm/kg).
  - larger molecules in urine (pathology) variably effects SG measurement.
  - isosthenuria may be a sequela of urinary tract or renal disease or may be physiological and related to water intake.
- normal plasma osmolality in adult horses ranges from 275 to 312 mOsm/kg.
  - urine osmolality of >300 mOsm/kg indicates ability of kidneys to concentrate urine and will be three to four times that of plasma.

- pathological urine samples:
  - direct measurement of urine osmolality more accurate than SG.
- reagent strips not appropriate for examining equine urine.
- horses usually have alkaline urine (pH between 7.5 and 9.0):
  - may become acidic following high-intensity exercise.
  - ammonia odour to urine sample suggests bacterial breakdown of urea.
  - diluted urine usually neutral to slightly acidic pH.
- proteinuria may occur with:
  - pyuria, bacteriuria and glomerular disease.
  - physiologically – after exercise.
  - glomerular function may be temporarily altered by stress, fever, seizures, extreme environmental temperature and venous congestion in the kidneys:
    - results in reversible proteinuria.
  - commercial reagent strips often yield false-positive results for protein when alkaline urine is tested or when urine SG exceeds 1.035.
  - semi-quantitative sulphosalicylic acid precipitation or a colourimetric assay should be used to quantify the urine protein.
  - urine protein:creatinine ratio (UP/UC) is helpful in distinguishing primary glomerular disease (UP/UC >3; usually >5) from primary tubular disease (UP/UC <3).
- normal horse urine should not contain glucose:
  - glucosuria (glycosuria) – blood glucose levels exceed 11 mmol/l (200 mg/dl).
    - hyperglycaemia producing glucosuria can be caused by:
      - stress, exercise, sepsis, PPID or diabetes mellitus.
      - dextrose-containing fluids or parenteral nutrition compounds.
      - alpha-2 adrenergic agonists and treatment with corticosteroids.
    - glucosuria without hyperglycaemia is usually associated with renal tubular dysfunction.

TABLE 3.3 Characteristics of normal equine urine	
pH	7.5–9.0 (concentrated feeds tend to acidify urine)
Specific gravity	1.018–1.025
Osmolality	~900 mOsm/kg
Glucose	negative
Protein	negative*
White blood cells	<5/hpf
Red blood cells	<5/hpf
Epithelial cells	none present if voided sample
Casts	usually negative – hyaline casts sometimes present
Crystals	common
Haemoglobin	negative
Myoglobin	negative
Bacteria	usually negative if catheterised urine

*False-positive protein result may occur on urine dipsticks with alkaline urine.

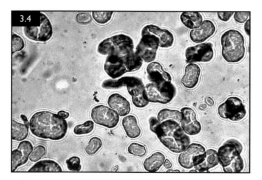

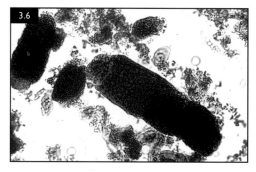

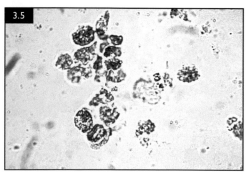

FIGS. 3.4, 3.5, 3.6 Urine sediment evaluation findings. Calcium carbonate crystals are a common component of equine urine (3.4). Leucocytes present in the urine (3.5). Granular casts (3.6). (Photographs courtesy R Jacobs.)

- urine sediment should be evaluated for cells, bacteria, casts and crystals (Figs. 3.4–3.6):
  - evaluation should be carried out no later than 1 hour after urine collection (Table 3.3).
  - urinary tract inflammation, infection, neoplasia or trauma may result in increased numbers of erythrocytes in the urine.
  - Pyuria (>5–8 white blood cells [WBCs]/hpf) is usually associated with urinary tract inflammation and/or infection.
  - bacteria can be present in the urine sediment in urogenital tract infection:
    - normal sediment contains no or few bacteria.

- ♦ note the absence of visible bacteria does not rule out infection.
  - ♦ quantitative and qualitative bacterial culture should be performed on urine that was collected by catheterisation or, in foals, by cystocentesis.
- casts are mucoproteineous substances that are formed within the distal renal tubules:
  - ○ cast formation increases with urinary tract inflammation or/and infection.
  - ○ rare in normal equine urine – usually dissolve in alkaline urine.
  - ○ present only transiently.
  - ○ may not be detected in all cases of acute renal disease or in every urine sample.
  - ○ **absence of casts does not therefore rule out renal disease.**
- crystals are usually abundant in equine urine:
  - ○ may interfere with urine sediment evaluation.
  - ○ calcium carbonate crystals most common followed by triple phosphate and, rarely, calcium oxalate.
- gamma-glutamyltransferase (GGT) found in high concentrations in epithelial cells lining the proximal renal tubules:
  - ○ physiologically, its activity in urine arises from cell turnover.
  - ○ any damage to the renal tubular epithelium will increase its activity in urine.

Haematology and serum chemical analysis

- Azotaemia:
  - ○ increase in serum urea and creatinine:
    - ♦ pre-renal, renal or post-renal disease.
    - ♦ pre-renal azotaemia most common form and associated with:
      - – dehydration.
      - – other disturbances causing decreased renal perfusion.
    - ♦ renal azotaemia is associated with intrinsic renal failure:
      - – functional loss of approximately 70% of nephrons.
    - ♦ post-renal azotaemia is associated with urinary tract obstruction or rupture.

- ○ identification of cause of azotaemia critical to management of urinary tract disease:
  - ♦ clinical signs and laboratory findings need to be assessed simultaneously.
  - ♦ urine SG should be >1.018 in pre-renal azotaemia:
    - – no evidence of proteinuria, enzymuria or cylindruria.
    - – Cylindruria – presence of cylindrical, cigar-shaped renal casts.
- alterations in plasma and serum electrolyte levels encountered in certain types of urinary tract disease:
  - ○ sodium loss with polyuric renal failure:
    - ♦ varying degrees of hyponatraemia
    - ♦ urinary tract disruption and/or uroperitoneum produce hyponatraemia through resorption of urine, which is lower in sodium than serum.
  - ○ disrupted body electrolyte homeostasis affects serum concentration of chloride:
    - ♦ heavily excreted in polyuric renal failure in horses.
  - ○ serum potassium can be normal or elevated in renal failure:
    - ♦ markedly elevated in cases of uroperitoneum.
  - ○ acute kidney failure (AKI) – excretion of phosphorous in urine is disrupted:
    - ♦ increase in serum concentration.
    - ♦ AKI may also result in hypocalcaemia.
  - ○ **note hypercalcaemia and hypophosphataemia are often found in chronic kidney disease (CKD).**
- serum albumin and globulin concentrations variably decrease in chronic renal diseases:
  - ○ albumin loss greater than globulin because of its low molecular weight.
  - ○ neoplasia, glomerulonephritis, pyelonephritis or amyloidosis:
    - ♦ serum globulin concentration may increase due to chronic antigen stimulation.
- mild to moderate anaemia may be associated with CKD:
  - ○ decreased erythropoietin production and a shortened erythrocyte lifespan.

Fractional excretion of electrolytes

- repeated collection of urine samples obtained on:
  - consecutive days at same time of day and same stage of daily routine.
  - plasma collected at the same time.
  - fractional excretion (FE) should be calculated using the following equation:

$$FE = (([Cr]_{plasma}/[Cr]_{urine}) \times ([X]_{plasma}/[X]_{urine})) \times 100$$

where $[Cr]_{plasma}$ and $[Cr]_{urine}$ are the creatinine (Cr) concentrations in the plasma and urine, respectively, and $[X]_{plasma}$ and $[X]_{urine}$ are the concentrations of a specific electrolyte or mineral in plasma and urine, respectively (Table 3.4).

TABLE 3.4 Reference intervals for fractional excretion of electrolytes	
ELECTROLYTE	RANGE (%)
Sodium	<1.00
Chloride	<1.50
Potassium	15–65
Phosphorous	<0.50
Calcium	<7
Magnesium	<15

- suggested reference intervals can be influenced by diet.
- concern about reliability of results has limited its use in the detection of renal disease.

Quantification of glomerular filtration rate (GFR)

- indicator of functional renal mass:
  - assesses excretion of endogenous or exogenous substances (inulin, creatinine, sodium sulphanilate) that are neither secreted nor resorbed after filtration into the tubular lumen.
  - use of radiopharmaceuticals (sensitive tool) only available in referral centres.

Water-deprivation test

- simple test that determines whether an apparent inability to concentrate urine is caused by psychogenic polydipsia or diabetes insipidus:
  - bladder emptied by catheterisation and a baseline urinalysis performed.
  - record baseline serum urea/creatinine levels and body weight before removal of food and water:
    - **should never be performed in a dehydrated or azotaemic horse.**
  - urine SG is measured 12 and 24 hours after the initiation of the test.
  - horses should be closely monitored during water deprivation:
    - if dehydration becomes apparent the test should be stopped:
      - 5% loss of body weight occurs, or the urine SG reaches 1.025.
  - SG reaches 1.025, then the ability to concentrate urine has been proven:
    - central or nephrogenic diabetes insipidus cases cannot concentrate urine.
    - horses with psychogenic polydipsia:
      - some cases, may not be able to concentrate urine because of washout of the medullar interstitial osmotic gradient:
      - such horses, a partial deprivation of water intake at 40 ml/kg/day should restore the osmotic gradient in a few days.

Diagnostic imaging

Ultrasonography

- kidneys relatively easily examined by transabdominal ultrasonography:
  - right kidney only imaged through dorsolateral extents of last three intercostal spaces (ICSs) (Fig. 3.7).
  - left kidney is imaged in the left paralumbar fossa (Fig. 3.8):
    - alternatively transrectally.
  - usually a 2.5- or 3-MHz probe:
    - 5-MHz probe sometimes adequate to examine the right kidney.
- AKI:
  - kidneys normal/increased in size, and corticomedullary junction may be indistinct.

3

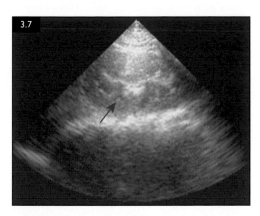

FIG. 3.7 Transabdominal ultrasonogram of the normal right kidney (arrow): long-axis cross-sectional view.

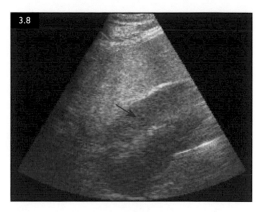

FIG. 3.8 Transabdominal ultrasonogram of the normal left kidney (arrow): long-axis cross-sectional view. The homogeneous organ above the kidney is the spleen.

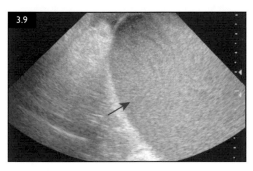

FIG. 3.9 Transabdominal ultrasonogram of the urinary bladder (arrow). A high concentration of calcium carbonate crystals and mucus makes equine urine quite echogenic.

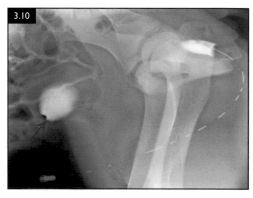

FIG. 3.10 Contrast cystogram. The urinary catheter (dotted line) extends throughout the urethra into the bladder (arrow).

- CKD:
  - kidneys are usually decreased in size with increased echogenicity.
  - cystic or mineralised areas can be associated with:
    - ◆ chronic renal diseases or, more often, with congenital abnormalities.
- calculi within the renal pelvis are occasionally seen.
- Doppler ultrasound may characterise renal haemodynamics.
- ultrasonography of the bladder is best performed transrectally in adults:
  - transabdominal approach may be useful in small horses and foals.
  - **note equine urine not homogeneous and has variable echogenicity** (Fig. 3.9).

  - bladder wall assessed for masses, both by ultrasonography and rectal palpation.
  - cystic calculi can be confirmed ultrasonographically.

Radiography

- only useful in foals and miniature horses.
- excretory urography or retrograde contrast studies (Fig. 3.10) may be performed.

Endoscopy

- useful diagnostic tool.
- flexible endoscope with an outside diameter of 12 mm or less should be used:
  - endoscopes less than 1 metre in length may not reach the bladder in males.

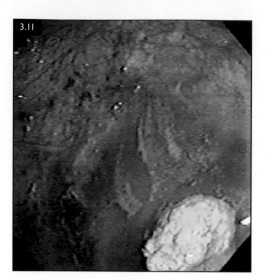

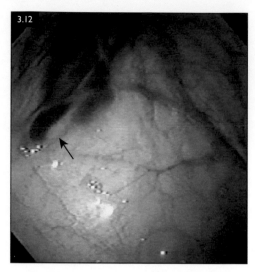

FIG. 3.11 Cystolith. Note the bladder mucosal ulceration.

FIG. 3.12 Ureteral opening. Note the bloody discharge (arrow).

- procedure is identical to that of bladder catheterisation:
  - urethra (especially ampullar portion), colliculus seminalis, bladder and ureteral openings should be examined.
  - inflammation, masses and calculi can be visualised (Fig. 3.11).
  - abnormal discharges from ureters (Fig. 3.12).
    - ◆ small amount of urine should enter bladder from ureteral openings at least every 60 seconds.
    - ◆ urine can be collected from each ureter.

- overinflation of the bladder causes patient discomfort and excessive straining:
  - **venous air embolism is rarely reported after room air insufflation and can be avoided if CO_2 is used to inflate the bladder.**

Renal biopsy

- rarely performed as only occasionally offers information more specific than that gathered by conservative diagnostic approaches and carries considerable risks.

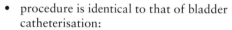

RENAL DISEASES

Acute kidney injury

Definition/overview

- sustained decrease in GFR leading to azotaemia and fluid and acid–base disturbances.
- caused by:
  - decreased renal perfusion (pre-renal or haemodynamic failure).
  - primary renal dysfunction (intrinsic renal failure).
  - obstruction of urine flow (post-renal failure).
  - pre-renal failure and renal failure are the most common.

Aetiology/pathophysiology

- any cause of renal hypoperfusion (especially prolonged) such as:
  - dehydration from GI disease, heavy exercise or blood loss (pre-renal).
- nephrotoxins often associated with intrinsic renal failure:
  - especially in horses with concurrent renal hypoperfusion.
  - aminoglycoside antimicrobials, tetracyclines, NSAIDs.
  - endogenous pigments (myoglobin or haemoglobin).
  - heavy metals.
  - vitamins D or K_3.

- plant toxins (onions, red maple leaves and, rarely, acorn poisoning).
- Glomerulonephritis, interstitial nephritis and renal microvascular thrombosis are more complex entities of intrinsic renal failure.
- post-renal failure may occur from functional or mechanical urinary tract obstruction or urinary tract rupture.

Clinical presentation

- most common are:
  - anorexia, dehydration and depression.
  - abdominal discomfort.
  - pigmenturia.
  - PU/PD.
- concurrent disease may result in additional clinical abnormalities.
- alterations in vital parameters depend on underlying disease causing the AKI:
  - uraemia can cause encephalopathy (uncommon).
  - urine production is variable:
    - anuric   - oliguric.
    - normouric or polyuric.
  - oedema may be present with anuric or oliguric renal failure.
  - mucous membranes are usually injected or hyperaemic.
- Laminitis may be present subsequent to AKI or the underlying disease process.

Differential diagnosis

- shock   - urinary tract calculi.
- sabulous urolithiasis.
- cystitis   - bladder paralysis.
- peritonitis   - visceral pain.
- Cantharidin toxicosis.

Diagnosis

- based on history, clinical signs, serum biochemical analysis and urinalysis.
- palpation per rectum should be performed where possible to assess:
  - renal size, the presence of perirenal oedema and renal pain:
    - typically, kidneys increased in size in AKI (Fig. 3.13).
    - decreased in CKD.
  - presence of obstruction in the ureters, bladder or urethra.
- serum biochemical analysis:

FIG. 3.13 Enlarged kidney in a stallion (Arab) with AKI. The sagittal length of this kidney is 19 cm (7.5 in.). The normal length of the kidney is comparable to the length of 2.5–3 vertebrae, which in this case should be approximately 12 cm (4.7 in.).

  - increases in blood urea and creatinine are invariably present.
  - hyponatraemia, hypochloraemia and hyperkalaemia.
  - hypocalcaemia can also be associated with AKI:
    - important to define 'true' hypocalcaemia.
    - decrease in ionised calcium – not total serum calcium.
  - phosphorous excretion in AKI is disrupted leading to increase in serum levels.
- **urinalysis**
  - essential to differentiate pre-renal from renal failure and characterise renal failure:
    - pre-renal failure, urine should be concentrated (SG >1.018).
    - isosthenuria (SG 1.008–1.012) present with renal failure.
  - mild to moderate proteinuria may be present:
    - depends on the aetiology and severity.
  - glomerular or tubular damage is present:
    - changes to urine sediment:
      - presence of casts and increased numbers of erythrocytes/leucocytes.
    - glucosuria without hyperglycaemia is strong evidence of renal tubular damage.
    - urinary GGT: urinary creatinine ratio of >25 suggestive of renal tubular disease but not specific.

- ◆ fractional clearance of sodium and the urine creatinine: serum creatinine ratio may be helpful.
- **ultrasonography**
  - ○ renal enlargement, perirenal oedema, loss of detail at the corticomedullary junction and nephroliths.
  - ○ dilation of renal pelvis may be evident with urinary outflow obstruction (Fig. 3.14):
    - ◆ renal biopsy has a limited diagnostic value.
    - ◆ urinary tract endoscopy may be useful if obstructive urinary tract disease is suspected.

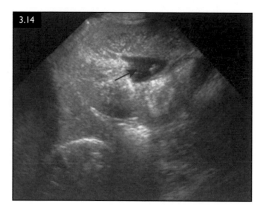

FIG. 3.14 Transabdominal ultrasonogram of the right kidney in a horse with suspected CKD. Note the dilation of the renal pelvis (arrow). Hydronephrosis is present on the dorsomedial part of the kidney.

Management

- primary disease should be managed accordingly.
- important to rule out urinary tract obstruction:
  - ○ especially in patients who present with oliguria or anuria.
- nephrotoxic drugs should be discontinued or, if treatment is necessary, the dosing regimen formulated at minimal possible effective dose.
- fluid therapy is essential for the treatment of AKI, regardless of the cause:
  - ○ diagnostic samples should be collected prior to fluid therapy if possible.
  - ○ restores fluid and acid–base deficits.
  - ○ prevent/reduce intrinsic renal lesions.

- ○ consider whether pre-renal, renal or post-renal failure and polyuria, normouria, oliguria or anuria are present:
  - ◆ oliguric or anuric patients: therapy should be conservative initially:
    - – set at 50% of the calculated requirements (estimated level of dehydration × body weight = amount of fluid required in litres).
    - – pulmonary sounds, body weight, central venous pressure and urine production should be monitored to prevent the development of overhydration and pulmonary oedema.
  - ◆ azotaemia should decrease rapidly following rehydration in cases of pre-renal failure.
- ○ hypernatraemia not present:
  - ◆ i/v administration of physiological saline (0.9% NaCl solution).
- ○ acutely hypernatraemic patients:
  - ◆ i/v 0.45% NaCl/2.5% dextrose solution.
  - ◆ slower volume correction prudent with hypernatraemia of longer/ unknown duration:
    - – reduce serum sodium concentration at a maximal rate of 0.5 mmol/l per hour or 10 mmol/l per day helps prevent cerebral oedema.
- ○ hyperkalaemia present (serum K^+ >5.5 mmol/l):
  - ◆ sodium bicarbonate given (1–2 mEq/kg i/v over 10–15 minutes).
  - ◆ alternatively, calcium borogluconate (0.5 ml/kg of 10% solution slowly i/v or added to 5 litres of fluids and infused over 1 hour).
  - ◆ both counteract the cardiotoxic effects of hyperkalaemia.
- ○ azotaemia begins to resolve:
  - ◆ fluid therapy should be continued at a rate that maintains normal hydration of the horse (maintenance rate of 60 ml/kg/day).
  - ◆ discontinued when:
    - – patient's mentation is normal.
    - – creatinine values less than 10–15% over high normal reference range.

- – other serum biochemical values are normal.
  - – discontinue fluid therapy gradually.
  - – monitor serum creatinine regularly following cessation.
  - ○ oliguria or anuria persists after rehydration – more aggressive therapy indicated:
    - ◆ furosemide (1–4 mg/kg i/v q6 h).
    - ◆ mannitol (0.25–1.0 g/kg i/v as 20% solution q4–6 h).
    - ◆ dopamine (120 mg in 1 litre of 0.9% NaCl or 5% dextrose):
      - – given at 12.5 ml per minute to achieve 3 µg/kg/min.
    - ◆ fluid therapy.
    - ◆ treatment with mannitol and dopamine only if close monitoring available.
- peritoneal dialysis sometimes helpful in relieving severe azotaemia but requires advanced equipment.
- surgical intervention may be indicated with urinary tract obstruction or rupture:
  - ○ stabilisation of horse with fluid therapy and gradual drainage of peritoneal fluid are indicated if uroperitoneum is present.
- during the later polyuric recovery phase of AKI:
  - ○ i/v or oral electrolyte/salt supplementation is required.
- most horses with AKI do not require specialised dietary support:
  - ○ feeding grass forage can provide a diet low in protein, phosphorous and calcium.
  - ○ concentrated feed (not necessary in most horses):
    - ◆ fed at no more than 0.6–0.7 kg/100 kg of body weight per day.

Prognosis

- affected by the duration of AKI prior to initiation of therapy.
- pre-renal failure is good if:
  - ○ primary disease can be controlled, and appropriate fluid therapy provided.
- degree of tubular and/or interstitial damage is always present in AKI.
- rapid resolution of azotaemia (decrease in urea of 25–50% within 24 hours):
  - ○ associated with a favourable prognosis.

- poor prognosis for recovery for horses that:
  - ○ oliguric for 48 hours or more before the initiation of therapy.
  - ○ develop complications such as:
    - ◆ generalised oedema, laminitis or encephalopathy.
  - ○ remain oliguric despite fluid therapy.
  - ○ many of these horses go into CKD.
- horses that recover from AKI are more prone to develop renal failure in the future.
- post-renal failure, the prognosis depends on the ability to correct the underlying problem and whether intrinsic renal failure has developed.

Specific aetiologies associated with AKI

Aminoglycoside toxicity

- common cause of intrinsic AKI in horses.
- neomycin most nephrotoxic of aminoglycosides and streptomycin the least:
  - ○ gentamicin most associated with renal failure because of widespread use.
- aminoglycosides accumulate within tubular epithelial cells, disrupt phospholipid metabolism, cause tubular necrosis and renal vasoconstriction:
  - ○ damage to tubular epithelial cells usually develops after 3–5 days administration.
  - ○ most common in dehydrated and/or hypotensive animals.
  - ○ concurrent treatment with NSAIDs, diuretics, cisplatin and/or cephalosporins may aggravate the disease.
- diagnosis is based on a history of aminoglycoside administration, clinical signs and a laboratory diagnosis of intrinsic AKI as discussed above.
- nephrotoxic drugs should be promptly discontinued in horses that show signs of AKI:
  - ○ recovery on discontinuing the drug and provision of supportive care in most cases.
- prevention requires several measures:
  - ○ ensure patients are adequately hydrated.

- o no underlying renal disease prior to using drug.
- o monitoring of peak and trough drug levels is very useful, where possible.

NSAIDs

- may lead to AKI if used at excessive doses or in dehydrated or hypotensive horses.
- cause medullary crest necrosis by disruption of renal synthesis of prostaglandins, which regulate renal perfusion.
- characteristic signs of renal failure plus haematuria may also be present.
- NSAID-associated AKI suspected – administration should be ceased:
  - o alternative analgesic drugs, including alpha-2 agonists and/or opioids, given parenterally, transdermally or via the epidural route.
  - o NSAIDs must be used – administer lowest possible doses:
    - ♦ COX-2 selective NSAID, such as firocoxib.
    - ♦ normal hydration and blood pressure must be maintained.
  - o phenylbutazone suggested to be the most nephrotoxic, then flunixin meglumine and ketoprofen.
- treatment as AKI.

Pigment nephropathy

- haemoglobin and myoglobin are potentially nephrotoxic.
- horses with severe haemolysis or rhabdomyolysis at risk of developing pigment nephropathy, particularly if dehydrated or hypotensive.
- nephropathy associated with direct tubular toxicity, tubular obstruction and renal vasoconstriction.
- coagulopathies associated with severe haemolysis may also affect renal vasculature integrity.
- identification and control of inciting cause are essential plus aggressive fluid therapy:
  - o sodium bicarbonate may be indicated when myoglobinuria associated with AKI:
    - ♦ alkalisation of urine increases urinary myoglobin solubility and reduces intrinsic damage to the kidney.

Miscellaneous drug and other toxicities

- variety of nephrotoxins, including cisplatin, proton pump inhibitors (omeprazole), vitamin K_3, vitamin D, tetracycline, polymyxin B, amphotericin B and heavy metals.
- ingestion of acorns, ochratoxins and cantharidin may induce renal failure.
- treatment involves removal of the initiating factor and general principles for the treatment of AKI:
  - o recently ingested:
    - ♦ stomach lavaged.
    - ♦ charcoal orally (1–3 g/kg via nasogastric tube).
  - o treatment with laxatives warranted after charcoal treatment.

Chronic kidney disease

Definition/overview

- irreversible loss of functional nephrons.
- clinical signs only occur when functioning nephrons decreases below 20–30%.
- all causes of AKI can lead to CKD.
- several other metabolic, immunological, infectious, obstructive and congenital disorders can cause CKD.

Aetiology/pathophysiology

- acquired disorders are the most common cause:
  - o extensive damage to the functional nephrons usually follows AKI:
    - ♦ prolonged renal hypoperfusion
    - ♦ exposure to nephrotoxins such as aminoglycosides, oxytetracycline, NSAIDs, endogenous pigments (myoglobin or haemoglobin), heavy metals, vitamin D or K_3 and plant toxins.
    - ♦ glomerulonephritis, interstitial nephritis, renal microvascular thrombosis, renal amyloidosis, renal pelvic calculi and renal neoplasia.
- CKD develops when tubular and glomerular damage exceeds renal reserve capacity:
  - o renal function decreases:
    - ♦ surviving nephrons undergo functional changes.

♦ permit horse to regulate water and solute homeostasis.
  ○ disease progression:
    ♦ compensatory mechanism may be overwhelmed.

Clinical presentation

- anorexia and weight loss are the most common.
- poor athletic performance may be detected early in the disease.
- depression and lethargy develop with progression of uraemia.
- rough hair coat, ventral oedema and PU/PD commonly associated with late CKD.
- urea may be converted to ammonia on mucosal surfaces of GI tract, resulting in ulceration.
- uraemic halitosis and excessive dental tartar formation may be present (Fig. 3.15).
- encephalopathy is a possible but uncommon sequela of uraemia.

Differential diagnosis

- pleuropneumonia   • peritonitis.
- malabsorption/maldigestion syndrome.
- neurological disorders   • neoplasia.
- ruptured bladder.
- renal tubular acidosis.

Diagnosis

- based on clinical signs, blood samples, urine analysis and ultrasonography:

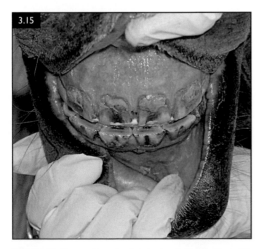

FIG. 3.15 Excessive dental tartar in a horse with CKD.

  ○ persistent increases in blood urea and creatinine.
  ○ normocytic, normochromic anaemia.
  ○ hypoalbuminaemia, +/– hypoproteinaemia
  ○ hyponatraemia, hypochloraemia, hypercalcaemia, hypophosphataemia and low plasma bicarbonate concentration are variably present.
  ○ excessive urinary losses of electrolytes can be expressed as metabolic acidosis or, less frequently, metabolic alkalosis.
- urine analysis:
  ○ persistent isosthenuria (SG 1.008–1.012) in presence of azotaemia or dehydration confirms presence of renal failure.
  ○ proteinuria is substantial:
    ♦ SG of urine may rise to, or exceed, 1.020.
    ♦ despite an inability of the kidneys to concentrate urine.
  ○ urine sediment free of cells/casts unless CKD associated with pyelonephritis:
    ♦ RBCs, leucocytes, casts and bacteria may be present.
    ♦ bacterial culture of urine (catheterised sample) in all suspected CKD cases.
- rectal palpation to evaluate the structure and size of the left kidney and ureters:
  ○ kidney is usually normal or small in size.
  ○ neoplasia, infection or urinary tract obstruction:
    ♦ kidney and/or ureters may be enlarged.
- ultrasonography:
  ○ kidneys smaller and hyperechogenic compared with normal.
  ○ loss of distinction of the corticomedullary junction (Fig. 3.16).
  ○ cysts or nephroliths may be visualised (Fig. 3.17).
- kidney biopsy may be helpful in the diagnosis of pyelonephritis or a congenital abnormality.

Management

- address any underlying disease.
- cease administration of nephrotoxic or potentially nephrotoxic drugs.
- intravenous fluid therapy is indicated in:

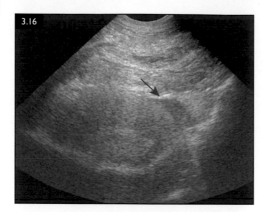

FIG. 3.16 CKD. Note complete loss of the cortico-medullary junction in the right kidney. The arrow indicates an area of mineralisation.

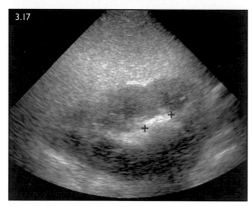

FIG. 3.17 Transabdominal ultrasonogram of the left kidney in a horse with suspected CKD. Hyperechogenic areas between the crosses indicate mineralised debris or the presence of a renal calculus.

- o acute exacerbation of CKD, azotaemic animals, treatment of underlying or concomitant disease or in dehydrated animals.
- o physiological saline (0.9% NaCl solution) is the fluid of choice:
  - ♦ balanced electrolyte solution used if moderate to severe hyperkalaemia not present.
  - ♦ replace fluid deficits plus maintenance requirements (65 ml/kg/day) and ongoing losses.
  - ♦ oliguria or anuria is present:
    - – close observation during fluid therapy.
    - – ensure overhydration and resultant oedema do not occur.
  - ♦ oliguria or anuria is uncommon:

- – furosemide, mannitol and/or dopamine are rarely indicated.
- • supportive management is essential:
  - o palatable diet low in protein, calcium and phosphorous:
    - ♦ high-quality grass forage, corn and oats.
    - ♦ supplementation with fat (high-fat pellets, rice bran, vegetable oil) if increased caloric intake desired.
    - ♦ legumes should be avoided because they are high in protein and calcium.
    - ♦ bran should be avoided, as it is high in protein and phosphorous.
  - o vitamin supplementation to compensate for excessive polyuria-induced losses of B vitamins.
  - o free access to water is critical:
    - ♦ supplementation with oral electrolytes:
      - – NaCl 25–50 g/day p/o.
      - – sodium bicarbonate 50–100 g/day p/o.
      - – potassium chloride (up to 50 g/day) if hypokalaemia develops.

Prognosis

- • long-term prognosis is grave:
  - o proper supportive care can affect the length and quality of life.
- • good management and regular monitoring of the disease progression may achieve a fair short-term prognosis for life.
- • athletic performance and breeding capabilities are limited.
- • poor with:
  - o anuria or oliguria.
  - o severe weight loss.
  - o severe elevations in blood urea and creatinine.
  - o azotaemia responds poorly to fluid therapy.

Specific aetiologies associated with chronic kidney disease

Congenital diseases of the kidney

- • Renal agenesis, hypoplasia and dysplasia are rare in horses.

- unilateral renal agenesis identified incidentally in mature horses:
  - renal function of remaining kidney is normal, but less renal reserve, so such horses are more prone to development of renal failure.
  - bilateral renal agenesis (not compatible with life) has been reported in a foal.
- renal hypoplasia when renal mass at least 50% smaller than normal:
  - renal failure develops if renal mass is <30% or if concurrent renal disease affects renal function.
- renal dysplasia is an abnormal differentiation of renal tissue that develops secondarily to *in utero* exposure to teratogens:
  - bilateral and unilateral disease have been reported.
- no specific treatment options for any of these conditions and long-term prognosis is poor.

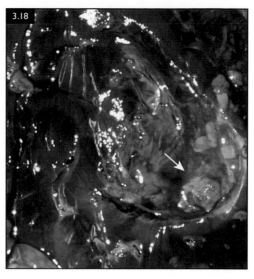

FIG. 3.18 Nephrolith 1 cm in diameter (arrow) in the right kidney.

Renal pelvis calculi

- Nephroliths (Fig. 3.18) develop within or adjacent to the renal pelvis:
  - can partially or completely obstruct the upper urinary tract:
    - ♦ passage of urine obstructed and hydronephrosis develops.
    - ♦ when obstruction is bilateral, CKD will develop.
- diagnosis is based on history and clinical signs:
  - unilateral disease, azotaemia may not be present:
    - ♦ urinalysis may reveal pigmenturia and/or microscopic haematuria.
  - transabdominal and transrectal ultrasonographic examination can identify nephroliths of significant size.
- successful dietary and/or medical procedures to dissolve nephroliths have not been reported in horses.
- unilateral nephrectomy is the treatment of choice if the remaining kidney is normal.

Interstitial nephritis

- common sequela of AKI.
- degree of damage and number of affected nephrons influences the severity of clinical signs and prognosis for short- and long-term recovery.

Immune-mediated glomerulonephritis

- intraglomerular inflammation and cellular proliferation associated with haematuria.
- antigen–antibody complexes are deposited in the glomeruli of kidneys causing a local inflammatory response and vasculitis:
  - most often seen following streptococcal infections.
  - circulating immune complexes of other chronic diseases such as leptospirosis, *Borrelia burgdorferi* and herpesvirus infections probably also lead to glomerular deposits.
- persistent deposition of complexes leads to irreversible damage and CKD.
- definitive diagnosis is via histopathological and immunofluorescence examination of renal biopsies or necropsy samples.
- treatment should aim at removal of the initiating cause of the glomerulonephritis and CKD treatment as described previously.

Pyelonephritis

Definition/overview

- suppurative bacterial infection of the kidney (Fig. 3.19).

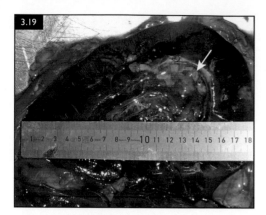

FIG. 3.19 Pyelonephritis and renal hyperaemia in the right kidney. Note the purulent debris in the renal medulla (arrow).

- uncommon cause of renal failure in the mature horse.
- often develops consequent to urolithiasis, trauma, neurogenic incontinence or bladder paralysis, which predisposes to ascending infection.

Aetiology/pathophysiology

- ascending infection from lower urinary tract is usual source of bacterial colonisation.
- **haematogenous spread of infection to the kidneys can occur, although rarely.**
- commonly implicated in ascending infections are *Corynebacterium* spp., *E. coli*, *Proteus mirabilis*, *Klebsiella* spp., *Enterobacter* spp., *Actinobacillus* spp., *Salmonella* spp., *Pseudomonas* spp. and *Streptococcus* spp.
- common agents for haematogenous infection are *Leptospira* spp., *Salmonella* spp., *Actinobacillus equuli* and *Streptococcus equi*.
- long-standing pyelonephritis leads to progressive damage of all renal structures.

Clinical presentation

- signs of a systemic disease:
  - weight loss, fever, PU/PD, generalised weakness and depression or lethargy.
- concurrent renal or post-renal failure may be present depending on the severity of renal damage and whether nephroliths or ureteroliths have formed and caused obstructions.

Diagnosis

- urinalysis is essential:
  - microscopic or macroscopic haematuria with pyuria.
  - bacteria may be evident microscopically:
    - **absence of visible bacteria does not rule out infection.**
  - urine culture must be performed, ideally from a catheterised urine sample.
  - urine is usually concentrated:
    - SG >1.020
    - unless renal failure is present (isosthenuria will be identified).
- blood samples:
  - neutrophilic leucocytosis often present, with increased plasma fibrinogen.
  - azotaemia not present unless renal failure has developed.
- ultrasonographic examination of the bladder and kidneys should be performed:
  - identification of nephroliths and changes in renal architecture (Fig. 3.20).
- palpation per rectum of kidneys.
- possible inciting causes evaluated.

Management

- appropriate antimicrobial treatment based on bacterial culture/sensitivity is essential:
  - beta-lactam antibiotics (procaine penicillin 20,000 IU/kg i/m q12 h or ceftiofur sodium 2.2 mg/kg i/m or i/v q12 h) initially until culture and sensitivity results.

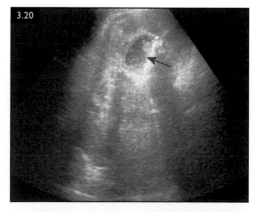

FIG. 3.20 Transabdominal ultrasonogram of the right kidney in a horse with pyelonephritis. Note the hyperechogenic debris located in the distended renal pelvis (arrow).

o repeated urinalyses should be performed to assess response to treatment:
  ♦ stop antimicrobial treatment 1 week after negative urine culture.
- free access to water provided:
  o add table salt (2 tablespoons q12–24 h) to encourage drinking and increase urine production.
- nephrectomy considered if disease is unilateral, severe and poorly responsive to medical therapy.

Prognosis

- long-term prognosis for survival of early cases is fair.
- pyelonephritis often diagnosed at an advanced stage, and therefore the prognosis is often poor.

Renal tumours

- disseminated tumours of any type, most often lymphoma and haemangiosarcoma, may localise to the kidney.
- primary renal neoplasia is rare in horses:
  o renal cell carcinoma (adenocarcinoma) in older horses.
  o nephroblastoma in young horses.
- treatment of choice for unilateral renal neoplasia is nephrectomy.
- prognosis for recovery can be grave.

Hydronephrosis

- obstruction of urine flow from the proximal urinary tract results in a progressive interstitial fibrosis and atrophy of the kidney.
- if unilateral, may remain clinically inapparent for a long period of time.
- dilation of the renal pelvis may be evident ultrasonographically.
- ureteral distension (Fig. 3.21) may be palpable per rectum or evident

FIG. 3.21 Distended right ureter (upper) in a horse with pyelonephritis and nephroliths. Note the purulent debris in the ureteral lumen. The left ureter (lower) is normal.

ultrasonographically depending on the location of the obstruction.
- nephrolithiasis is the most common cause of hydronephrosis:
  o other causes include renal and bladder neoplasia, cystitis, acquired strictures of the urethra or any inflammatory condition that surrounds the urinary tract.
- treatment should be directed at correcting the primary cause of the disease.
- prolonged or repeated periods of obstruction cause irreversible renal damage and CKD.

Amyloidosis

- rarely reported:
  o horses used for antiserum production or following chronic infection.
- amyloid deposition disrupts tissue architecture of the kidney and direct toxicity of amyloidogenic precursor proteins cause renal disease:
  o disturb normal renal function and lead to end-stage renal disease.
- no effective treatments reported in horses.

RENAL TUBULAR DISORDERS

Renal tubular acidosis (RTA)

Definition/overview

- uncommon condition in which renal tubules are unable to acidify urine.

- results in a continued state of metabolic acidosis:
  o urine pH remains, however, neutral or alkaline.

Aetiology/pathophysiology

- probably a secondary condition.
- two types of RTA have been reported in horses:
  - type I (distal tubular acidosis):
    - unable to acidify urine because of inadequate hydrogen ion secretion in distal renal tubules.
  - type II (proximal tubular acidosis):
    - inability of proximal renal tubules to resorb bicarbonate, which is subsequently lost in urine.
    - often a self-limiting disease.
- hyperchloraemia and hypokalaemia occur concurrently.

Clinical presentation

- anorexia, weight loss, depression and weakness are main presenting complaints.
- ataxia, poor performance, ill-thrift, tachypnoea and tachycardia also reported.

Differential diagnosis

- renal failure
- uroperitoneum.
- renal calculi
- neurological disorders.
- malabsorption/maldigestion syndrome.
- exertion.

Diagnosis

- electrolyte and acid–base disturbances are the main haematological abnormalities:
  - severe hyperchloraemic metabolic acidosis with a low strong ion difference.
  - hyponatraemia and/or hypokalaemia may also be present.
  - high fractional excretion of sodium in type I RTA.
  - low fractional excretion of potassium in type II RTA.
- blood urea nitrogen and creatinine should be normal unless dehydration is present:
  - if elevated indicates concurrent renal disease.
- urine pH is:
  - neutral to alkaline in type I RTA.
  - alkaline to acidic in type II RTA.
- ammonium chloride loading has been used to detect type I RTA.
- type II RTA is based on clinical and laboratory findings, and exclusion of type I RTA.

Management

- treatment and prognosis of both types of RTA are similar.
- i/v administration of sodium bicarbonate to correct the metabolic acidosis:
  - initial treatment administered gradually to replace the estimated bicarbonate deficit:
    - $0.3 \times$ body weight [kg] $\times$ base deficit = bicarbonate deficit in mmol/l.
    - return plasma bicarbonate concentration to values above 20 mEq/l and blood pH above 7.3.
  - once stabilised:
    - controlled by oral administration of sodium bicarbonate (50–150 g q12–24 h).
    - oral potassium supplementation often necessary during initial stages of treatment.
    - serum electrolyte levels and blood gases should be monitored regularly.

Prognosis

- based on severity of underlying renal disorder, and duration of response to initial therapy.
- short-term prognosis is usually good with proper treatment.
- relapses are common if renal disease is present.

Diabetes insipidus

Definition/overview

- uncommon cause of PU/PD in horses.

Aetiology/pathophysiology

- inadequate antidiuretic hormone (ADH) (vasopressin) is produced:
  - neurogenic or central diabetes insipidus.
  - can develop secondary to head trauma, encephalomyelitis or PPID.
- distal tubules, collecting tubules and collecting ducts are unable to respond to ADH:
  - nephrogenic diabetes insipidus.
  - secondary to many types of renal disease, especially renal medulla damage.
- hereditary basis to the disease is possible.

Clinical presentation

- PU/PD should be the sole presenting complaint.

Differential diagnosis

- includes CKD, psychogenic polydipsia, diabetes mellitus and PPID.

Diagnosis

- physical examination is unremarkable.
- urinalysis should be normal apart from a lack of concentration of urine.
- blood urea and creatinine levels are normal unless dehydration is present.
- water-deprivation testing should be performed as described earlier:
  - **never performed in a dehydrated or azotaemic horse.**
  - **closely monitored during the test to avoid severe hypertonic dehydration.**
  - inability to concentrate urine during test:
    - ◆ psychogenic polydipsia with medullary interstitial osmotic gradient cannot be ruled out initially in horses not responding to water deprivation.
    - ◆ partial deprivation of water intake at 40 ml/kg/day should be performed prior to repetition of the water-deprivation test:
      - – urine concentration still does not occur in diabetes insipidus.
- alternative diagnostic methods include:
  - infusion of hypertonic saline in normal horses should stimulate urine concentration across renal tubules.
  - ADH (vasopressin) challenge should stimulate urine concentration and is used to differentiate nephrogenic from central diabetes insipidus.

Management

- secondary diabetes insipidus should be managed via treatment of the primary disease.
- successful treatment of primary or idiopathic diabetes insipidus has not been reported.

Prognosis

- fair if not associated with underlying renal or neurological disease and access to water is always available:
  - prone to dehydration if water is restricted.
- secondary diabetes insipidus depends on the prognosis for the primary disease.

DISEASES OF THE URETERS

Ectopic ureters

- rare deformation in horses and most often reported in fillies.
- usually noted in foals with a complaint of persistent urine dribbling and perineal dermatitis (urine scalding).
- no other clinical or haematological abnormalities.
- endoscopic examination of the vagina and distal urinary tract may reveal the orifice of the ectopic ureter, but visualisation is often difficult.
- excretory urogram is useful in diagnosis, especially in young animals.
- surgical correction may be required:
  - depends on location of ectopic ureter, severity of clinical signs and intended use of animal.
  - relocation of the ectopic ureter into the bladder may be possible.
  - nephrectomy can be performed in patients with unilateral disease.

Ureterolithiasis

Definition/overview

- presence of calculi in one or both ureters, a rare problem in the horse.

Aetiology/pathophysiology

- sequela to degenerative or inflammatory processes in the kidney.
- inflammatory debris can serve as a nidus for calculus formation (see Fig. 3.21) within the ureter.
- nephroliths may move into the ureters and cause obstruction.

Clinical presentation

- unilateral disease:
  - low-grade intermittent colic may be only clinical sign.
- advanced cases:
  - signs consistent with CKD (develops secondarily to ureteral obstruction).

Diagnosis

- haematology is unremarkable until renal failure has developed.
- intermittent or persistent pigmenturia:
  - increased numbers of erythrocytes on urinalysis.
- SG variable depending on presence of renal failure and unilateral or bilateral disease:
  - bilateral obstructive disease, urine may not be obtained.
- ureterolith or enlarged ureter(s) may be palpable per rectum or evident ultrasonographically.
- urine culture should be performed but with variable results.

- multifocal urolithiasis elsewhere in the urinary tract is not uncommon.

Differential diagnosis

- calculi in other parts of the urinary tract.
- bladder paralysis • urinary tract trauma.
- renal failure.
- sabulous urolithiasis   • neoplasia.

Management

- surgical removal via a ureteral incision.
- unilateral nephrectomy in some cases.
- treatment of CKD is discussed elsewhere (see page 45).

Prognosis

- bilateral ureterolithiasis with CKD carries a grave prognosis.
- unilateral disease has a better prognosis:
  - especially if calculi can be removed successfully.

DISEASES OF THE URINARY BLADDER

Bacterial cystitis

Definition/overview

- inflammation of the bladder caused by bacterial infection:
  - rarely a primary disease.
- characterised by:
  - dysuria, stranguria and pollakiuria.
  - presence of blood, inflammatory cells and bacteria in the urine.

Aetiology/pathophysiology

- often secondary disease that can develop from:
  - urine stasis (bladder paralysis).
  - urinary tract catheterisation or trauma.
  - cystic calculi or neoplasia.
- *E. coli*, *Proteus* spp., *Pseudomonas* spp., *Klebsiella* spp., *Enterobacter* spp., *Streptococcus* spp. and *Staphylococcus* spp. are the most identified pathogens.
- mares are predisposed to development of cystitis.

Clinical presentation

- dysuria, stranguria and pollakiuria are the most common presenting complaints.
- signs of generalised disease such as fever, depression or weight loss should not be present with uncomplicated cystitis.
- urine scalding may be observed on the perineum or hindlimbs.

Differential diagnosis

- urolithiasis   • bladder paralysis.
- neoplasia.
- renal failure   • colic   • pyelonephritis.

Diagnosis

- physical examination findings typically non-specific.
- haematology usually unremarkable.
- urinalysis is diagnostic:
  - pyuria (more than 5 WBCs/hpf).
  - bacteria may be evident on cytological analysis of urine sediment:
    - absence does not rule out an infectious cause.

- o microscopic or macroscopic haematuria may be present.
- o usually concentrated urine (SG >1.020).
- o bacterial culture on a catheterised sample is preferred:
  - ♦ quantitative culture identification of >10,000 colony-forming units (CFU)/ml in a catheterised sample indicates infection.
- palpation per rectum to evaluate the bladder wall:
  - o determine whether uroliths may be present.
- endoscopy of the bladder via cystoscopy or transrectal ultrasonography may be helpful for investigation of primary disease.

Management

- identify and treat the initiating cause, if possible.
- use of an appropriate antimicrobial agent based on:
  - o sensitivity of isolated bacteria, and drug pharmacokinetics in the urinary tract.
  - o procaine penicillin (20,000 IU/kg i/m q12 h).
  - o trimethoprim/sulphadiazine (30 mg/kg p/o q12 h):
    - ♦ **do not use if renal failure accompanies cystitis.**
  - o other antimicrobials should be reserved for resistant infections.
- repeat urinalysis to assess response to treatment:
  - o antibiotic treatment up until 1 week after urine bacterial culture is negative.
- free access to water is essential:
  - o increase water intake by adding 50 g of table salt daily to the diet.
- bladder irrigation (0.9% NaCl) is beneficial in cases with cystic calculi or excessive amount of sediment:
  - o using endoscopy helps to minimise traumatic bladder irritation.

Prognosis

- good for primary cystitis.
- relapses are common.
- chronic cystitis, recurrent cystitis, ascending infection into the proximal urinary tract and neoplasia have a less favourable prognosis.

Bladder tumours

Definition/overview

- rarely diagnosed in horses and associated with a very poor prognosis.

Aetiology/pathophysiology

- usually older horses:
  - o squamous cell carcinoma most common followed by transitional cell carcinoma.
  - o metastases from other common neoplasias occasionally involve the bladder.
  - o fibromatous polyps are more common in younger horses.

Clinical presentation

- signs of disease are not usually evident until it is well advanced.
- weight loss and weakness most common presenting complaints.
- advanced stages are associated with loss of appetite, depression and lethargy.
- pollakiuria, stranguria and haematuria may be observed.

Differential diagnosis

- urolithiasis
- cystitis.
- renal failure
- paralysis.
- urinary tract trauma
- colic.
- neoplasia of other organ systems.

Diagnosis

- non-specific physical findings.
- urinalysis may reveal macroscopic or microscopic haematuria:
  - o neoplastic cells are often observed in sediment.
- bladder palpated per rectum:
  - o empty bladder prior to evaluation.
  - o thickened, irregular bladder wall or obvious mass may be palpable.
- transrectal and/or transabdominal ultrasonography can further evaluate the bladder.
- cystoscopy can be used to evaluate the bladder mucosa and obtain a biopsy.
- blood samples may reveal anaemia and hypoproteinaemia.

Management

- surgical excision and chemotherapy are variably successful.
- carcinomas are locally very invasive and may metastasise to other organs.
- prognosis is grave.

Cystic calculi

Definition/overview

- most common uroliths in the horse and usually identified in adults.

Aetiology/pathophysiology

- nidus in the form of organic debris needed as a base for calculus formation.
- risk factors for the development of cystic calculi not well understood:
  - tissue damage, cystitis, remaining suture material, supersaturation of urine with certain minerals and urine stasis may predispose to calculus development.
  - genetic predisposition is possible.
- mainly composed of calcium carbonate crystals:
  - mixed with calcium phosphate crystals
    - calculus becomes stronger.
  - most calculi are sphere-shaped stones.
- accumulation of crystalloid sludge (sabulous urolithiasis) can also occur (Fig. 3.22):
  - usually associated with bladder paralysis and urine stasis.

Clinical presentation

- stranguria, pollakiuria and haematuria are the most common presenting complaints:
  - haematuria may be more pronounced following exercise.
- signs of systemic disease such as fever, depression or anorexia not present.
- other signs include tenesmus, colic, incontinence and urine scalding.

Differential diagnosis

- cystitis
- oestrus
- pyelonephritis.
- bladder rupture
- neoplasia.
- calculi in other parts of the urinary tract.
- renal failure.

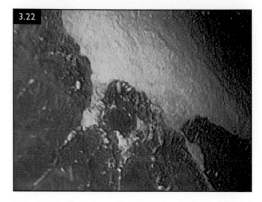

FIG. 3.22 Sabulous urolithiasis (yellow debris) associated with cystitis.

Diagnosis

- urinalysis:
  - urine should be concentrated.
  - proteinuria, microscopic haematuria and pyuria are common.
  - catheterised urine sample submitted for bacteriological culture:
    - >10,000 CFU/ml indicates concurrent urinary tract infection.
- haematology and serum biochemistry:
  - document accompanying problems of the proximal urinary tract.
  - no haematological abnormalities should be present with uncomplicated cases.
- calculi and/or a thickened bladder wall may be palpable per rectum:
  - concurrent bladder-wall inflammation may be painful.
  - Sabulous urolithiasis consists of an accumulation of sand-like debris in the bladder that may feel 'doughy' on palpation.
  - ultrasonographic examination of the urinary tract performed to exclude presence of calculi in other locations.
- cystoscopy helps further evaluate cystoliths (Fig. 3.23) or sabulous urolithiasis.

Management

- several techniques have been described for removal of cystic calculi:
  - laparocystotomy or laparoscopic cystotomy are commonly performed (Fig. 3.24).

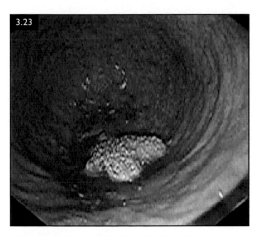

FIG. 3.23 Solitary cystolith in the bladder of a horse with dysuria. Note the roughened surface of the cystolith and the bloody urine.

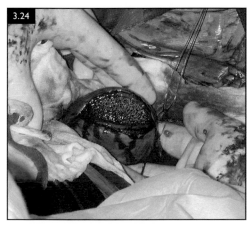

FIG. 3.24 Laparocystotomy: surgical removal of a cystolith. (Photo courtesy A Cruz.)

- subischial urethrotomy can be performed in a standing horse:
  - stricture or diverticulum formation at the incision site is a disadvantage.
- manual distension of the urethra in mares may permit removal of small stones.
- fragmentation and removal of cystic calculi by forceps in the bladder or lithotripsy (electrohydraulic, ballistic shock or laser) have been reported.
- pararectal cystotomy can aid successful removal of calculi, although there are significant postoperative complications.

- antimicrobial therapy, as described for cystitis, should be initiated if bacterial infection is present on urinalysis and urine culture, and following calculi removal.
- bladders containing sabulous uroliths should be irrigated with large volumes of fluid combined with a perineal urethrotomy as the most effective way to remove the uroliths.
- oral administration of 50–75 g of table salt once or twice a day may increase water intake/diuresis and is a useful preventive measure.
- free access to water should be provided.
- remove high-calcium feeds (alfalfa hay) from the diet.

Prognosis

- approximately 41% of horses have recurrence of cystolithiasis after treatment:
  - depends on thoroughness of calculus removal from the bladder and whether an underlying lesion is present.
- sabulous cystic deposits are mostly associated with bladder paralysis and the response to treatment is poor.

Bladder paralysis/ Neurogenic incontinence

Definition/overview

- incontinence develops as intravesicular pressure exceeds resting urethral pressure.
- incontinence and dysfunction of bladder control most often associated with neurological disorders in the central or peripheral nervous systems, and rarely with myogenic dysfunction in the bladder wall.

Aetiology/pathophysiology

- damage to the sacral spinal cord, pelvic and/or pudendal nerves leads to lower motor neuron (LMN) deficits:
  - equine herpesvirus (EHV)-1 myelitis. Equine protozoal meningitis/ myeloencephalopathy (EPM).
  - cauda equina syndrome, lumbosacral trauma and neoplasia.
  - sorghum toxicosis.

- ○ arboviral encephalomyelitis.
- ○ iatrogenic:
  - ◆ epidural administration of various pharmaceuticals.
  - ◆ after illegal tail altering procedures in Quarter Horses.
- ○ mares at increased risk for the development of LMN neurogenic incontinence from trauma during breeding and parturition.
- upper motor neuron (UMN) bladder dysfunction is associated with:
  - ○ damage to the supra-sacral spinal cord or/and brainstem.
  - ○ micturition is disabled via exaggerated urethral sphincter tone, despite the presence of a full bladder.
  - ○ chronic UMN lesions may, through the sacral spinal reflexes, allow partial voiding of urine.
- myogenic problems are rare, have been reported in geldings:
  - ○ lack a specific identifiable cause.

Clinical presentation

- dribbling of urine and urine scalding of the perineum (mares) and medial aspect of the hindlimbs (males and females).
- frequently posture to urinate and void little or no urine.
- clinical signs of underlying neurological diseases may be evident:
  - ○ UMN disorders frequently associated with recumbency and myopathy:
    - ◆ bladder distension leading to abdominal pain or frequent posturing to urinate.
    - ◆ bladder rupture may occur.
    - ◆ loss of anal and tail tone, faecal retention and perineal sensory deficits
    - ◆ hindlimb weakness and muscle atrophy, ataxia and penile prolapse.
    - ◆ haematuria if secondary infection or urolithiasis has developed.
- accumulation of large amounts of sabulous or mucoid urinary sediment:
  - ○ myogenic bladder dysfunction.
  - ○ less often, in LMN disease.
- severe and chronic dysfunction of the bladder wall can become permanent.

Differential diagnosis

- urolithiasis • cystitis • neoplasia.
- renal failure.
- cantharidin toxicosis • ectopic ureter.
- various neurological diseases.

Diagnosis

- detailed neurological evaluation essential to localise lesion and identify primary cause.
- per rectum examination may help classify the problem:
  - ○ LMN bladder is flaccid and easily expressible.
  - ○ exaggerated sphincter tone in a UMN bladder results in firm distension.
  - ○ sabulous urolithiasis may be palpable.
- transrectal and/or transabdominal ultrasonography and cystoscopy are helpful for eliminating other causes of incontinence and urine dribbling.
- haematology is unremarkable:
  - ○ unless bladder rupture or secondary upper urinary tract infection is present.
- urinalysis is normal unless secondary infection or urolithiasis has developed.
- urine culture performed in all cases.

Management

- manage underlying disease.
- regularly evacuate the bladder to prevent exacerbation of bladder atony and development of sabulous urolithiasis.
- provide nursing care, including daily cleaning of the perineum and hindlimbs, to reduce skin irritation.
- prophylactic antimicrobial treatment is indicated:
  - ○ recumbent animals.
  - ○ urinary tract infection suspected.
  - ○ frequent urinary catheterisation required.
- Phenoxybenzamine (0.7 mg/kg p/o q6 h) used in cases of UMN disease to decrease urethral sphincter tone although effectiveness is unclear.
- Bethanecol chloride (0.025−0.075 mg/ kg s/c or 0.2−0.4 mg/kg p/o q8 h) can be administered to improve detrusor muscle tone and strengthen bladder contraction:
  - ○ response can be poor with long-standing disease.

- o discontinued if no response within 3–5 days.
- Acepromazine (0.02–0.05 mg/kg i/m q8 h) and diazepam (0.02–0.1 mg/kg; slow i/v administration) may decrease urethral tone and help to void urine.

Urethral trauma and urethral defects

Definition/overview

- increasingly recognised in male horses.
- range of signs from haematuria to breeding disability and urinary tract obstruction.

Aetiology/pathophysiology

- male:
  - o trauma to the penis (Fig. 3.25).
  - o urethral calculi.
  - o masturbation control devices (stallion rings).
  - o post-surgical scar tissue.
  - o endoscopy of distal urinary tract.
  - o prolonged and traumatic urinary catheterisation.
- female:
  - o breeding injuries and dystocia.
- tears of the proximal urethra at the level of the ischial arch of the male more significant:

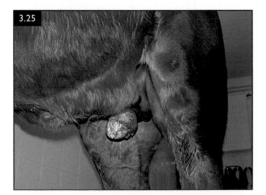

FIG. 3.25 Penile trauma. This stallion injured himself jumping over a fence with an erect penis. Note the oedema around the genitals, which may also be caused by urethral rupture and urine leakage.

- o may be result of corpus spongiosum penis damage due to dramatic pressure changes during ejaculation.

Prognosis

- guarded:
  - o depends on ability to treat the primary disease.
  - o prevent complications such as urinary tract infection or sabulous urolithiasis.

Clinical presentation

- haematuria at the end of urination.
- haemospermia in stallions.
- pollakiuria in some cases.
- penile, vaginal or perineal trauma may be apparent.

Differential diagnosis

- urolithiasis, urethritis, cystitis, bladder paralysis, neoplasia and sabulous urolithiasis.

Diagnosis

- no obvious clinical signs except for haematuria.
- examination of genitalia for signs of trauma:
  - o penis should be extruded, examined visually and carefully palpated.
- palpate the bladder per rectum.
- cystoscopy and urethroscopy carried out to confirm the urethral lesion (Fig. 3.26):
  - o **be careful not to exacerbate urethral lesion during urethroscopy.**
  - o retrograde urethrogram can be performed (Fig. 3.27) if the patency of the penile urethra is in question.
- ultrasonography can be useful in examination of surrounding tissue for any evidence of foreign bodies, scars or haematomas.
- urinalysis usually reveals haematuria only.
- prolonged haematuria may lead to severe anaemia.

Management

- most minor lesions to the urethra resolve spontaneously.

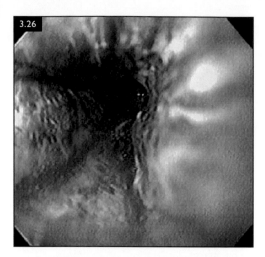

FIG. 3.28 Urethrolith (arrowed) in the urethral orifice. Urethrolithiasis is usually associated with urinary calculi in the bladder or proximal urinary tract. (Photo courtesy VK Kos.)

FIG. 3.26 On urethroscopy severe urethritis and mucosal erosions are present throughout the urethra.

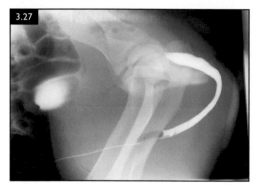

FIG. 3.27 Contrast urethrogram in a foal. Note the contrast material starting from the tip of the catheter and following the lumen of the urethra and bladder.

- lesions communicate with adjacent corpus spongiosum penis:
  - spontaneous resolution less likely.
  - ischial urethrotomy (circumvents intraurethral bleeding) allows healing.
- topical wound therapy:
  - systemic antimicrobials administered:
  - infection suspected or invasive therapeutic procedures.
  - procaine penicillin or trimethoprim/sulphadiazine.
- anti-inflammatory treatment may be necessary.

Prognosis

- varies depending on the severity of the urethral defect.

- severe trauma may obstruct urine flow primarily or secondarily with scar tissue formation and urethral stricture:
  - urethrotomy may be necessary to bypass the stricture.

Urethrolithiasis

Definition/overview

- urethral calculi develop mostly in male horses – uncommon in females.
- calculi are flushed from the bladder and lodge in the urethra (Fig. 3.28).
- outcome depends on degree of trauma to urethra and surrounding tissues.

Aetiology/pathophysiology

- most calculi initially lodge where the urethra narrows over the ischial arch.
- may move more distally and completely obstruct the urethra:
  - signs of colic.
  - if not treated, bladder rupture, uroperitoneum and post-renal AKI may develop.

Clinical presentation

- severity of clinical signs depends on whether complete urethral obstruction is present.
- frequent posturing to urinate with no or limited urine voided.
- pollakiuria and stranguria (Fig. 3.29).

- non-specific signs of abdominal pain:
  - complete obstruction, severe abdominal pain as progressive bladder distension.
- blood may be seen at the end of the urethral orifice.
- signs of uroperitoneum develop if the bladder ruptures.

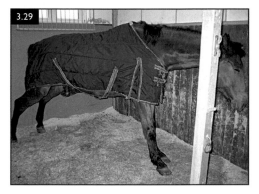

FIG. 3.29 Stranguria in a horse with urethrolithiasis.

Differential diagnosis

- urethral trauma   • neoplasia.
- urethritis   • cystitis.
- sabulous urolithiasis.
- bladder paralysis   • colic.

Diagnosis

- extend and carefully palpate the penis from tip to anus:
  - blood may occasionally be seen on the end of the urethra.
  - urethroliths may be palpable, depending on the location.
- palpate the bladder per rectum to assess its size:
  - often turgid and distended.
  - colic signs necessitate a thorough palpation of the rest of the abdomen.
- inability to pass a urinary catheter is suggestive of urethral obstruction:
  - urethral spasm can also inhibit advancement.
- urethroscopy usually provides a definitive diagnosis.
- urinalysis, if urine can be obtained, is consistent with signs of post-renal AKI:
  - urine bacterial culture should be performed.

- initial blood results usually unremarkable:
  - with bladder rupture, acid–base alterations and azotaemia ensue.
- **examine rest of the urinary tract for presence of other uroliths.**

Management

- calculi present in the distal urethra may be removed with haemostats (Fig. 3.30).
- calculi lodged further up the urethra can be removed via a urethrostomy:
  - lodged at ischial arch removed through a perineal urethrostomy (Fig. 3.31):
    - calculi can be crushed and then removed from the urethra.
    - iatrogenic damage can be sustained by the urethra and bladder.

FIG. 3.30 Removal of the urethrolith with a haemostat. (Photo courtesy VK Kos.)

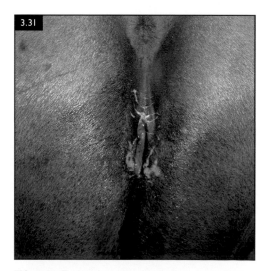

FIG. 3.31 Perineal urethrostomy site in a stallion with urethral obstruction and subsequent bladder rupture.

    o calculi lodged in less accessible parts of the urethra may require a urethrotomy performed under general anaesthesia.
- antimicrobials (procaine penicillin or trimethoprim/sulfadiazine) are necessary:
  - where therapeutic procedures are invasive.
  - concurrent infection is present:
    - based on urine culture and sensitivity results.

Prognosis

- depends on the severity of the urethral lesion and secondary complications.
- excessive tissue trauma increases the risk of urethral stricture.

Urethritis

Definition/overview

- inflammatory condition of the urethra that can be infectious or traumatic in origin.
- usually develops secondary to cystitis, urethral trauma, calculi or accessory gland infection.

Aetiology/pathophysiology

- favourable environment for colonisation by pathogens if defence mechanisms are overwhelmed.
- Gram-negative organisms predominant.
- *Candida* infection may occur in foals that undergo intensive antimicrobial therapy.
- *Habronema megastoma* may invade the urethral process, causing granulomas (Fig. 3.32).

Clinical presentation

- haematuria, haemospermia and stranguria are common.
- resentment of manual manipulation of the penis and sheath may be observed.

Differential diagnosis

- urethrolithiasis
- bladder paralysis.
- cantharidin toxicosis
- vaginitis.
- urethral trauma.
- neoplasia.

FIG. 3.32 *Habronema* granuloma (left) and normal urethral process (right). Redness of the urethral process was caused by catheterisation (right).

Diagnosis

- demonstration of the lesions by palpation and endoscopic examination.
- ultrasonography may be helpful in excluding involvement of the accessory sex glands:
  - fractionation/examination of the ejaculate may provide similar information.
- bacterial culture of a urethral swab, urine and semen should be performed.
- biopsy if habronemiasis suspected.

Management

- sheath cleaning and topical treatment with anti-inflammatory and antibacterial agents.
- any primary disease should be managed accordingly.
- systemic antimicrobials are occasionally indicated.
- in severe cases, oil-based antibiotic preparations can be infused via a urinary catheter into the pelvic urethra or used topically on the urethral process.
- *Habronema* granulomas managed with local moxidectin application, gentamycin and dexamethasone and, in advanced cases, surgical excision.

Prognosis

- may result in fibrous strictures, which carries less favourable prognosis.
- permanent damage to the urethra, may lead to recurrences and chronic urethritis.

GENERAL URINARY SYSTEM DISORDERS

3

Uroperitoneum

Definition/overview

- presence of free urine in the abdominal cavity.
- most recognised in foals between 24 and 48 hours old.
- male and septic foals are more likely to be affected.
- rupture of the bladder is most common.
- urinary tract rupture and urine leakage into abdominal cavity due to urinary calculi can affect adult horses.
- continuity of the urinary tract must be restored, or metabolic abnormalities caused by uroperitoneum will be fatal for the animal.

Aetiology/pathophysiology

- foals:
  - long and narrow urethra in colts resists the high pressure that is put on the bladder during parturition.
  - predisposes the weak bladder wall of the neonate to rupture.
  - urachal infection may also predispose to uroperitoneum.
- adult horses:
  - bladder rupture develops secondary to urethral obstruction, trauma, urinary catheterisation and, in mares, during dystocia.
- rupture of the bladder, urachus, ureter or renal pelvis may result in leakage of urine into the peritoneal cavity:
  - urine accumulation leads to azotaemia, hyperkalaemia, hyponatraemia, hypochloraemia and metabolic acidosis.
  - equilibration of urine electrolytes and water across the peritoneal membrane:
    - loss of sodium and chloride into the abdominal fluid/urine.
    - retention of potassium by diffusion from the abdominal fluid/urine.
    - urea readily diffuses across the peritoneal surface.
    - creatinine diffuses much more slowly.
    - urine causes a chemical peritonitis.

Clinical presentation

- abdominal discomfort and distension (Fig. 3.33).
- straining to urinate with little to no urine passed (Fig. 3.34):
  - voiding of small amount of urine does not exclude urinary tract rupture and uroperitoneum.
- fluid wave may be felt, or a sloshing sound heard on percussion of the abdomen.
- acid–base disturbance produces depression, anorexia, tachycardia and tachypnoea.
- respiratory distress can develop with severe abdominal distension, particularly in foals.

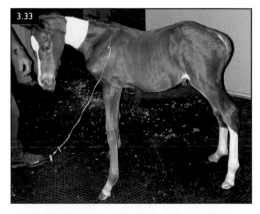

FIG. 3.33 Abdominal distension in a foal with uroperitoneum.

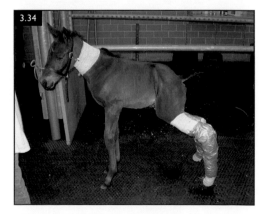

FIG. 3.34 A foal with uroperitoneum strains to urinate.

- shock and collapse will develop if not treated:
  - severe tachycardia may be present with severe hyperkalaemia.
- foals may also show signs of concurrent infection and sepsis:
  - fever, weakness, injected mucous membranes, diarrhoea and septic arthritis.
  - severe electrolyte disturbances can cause neurological abnormalities.

Differential diagnosis

- colic   • pleuropneumonia   • sepsis.
- endotoxaemia   • renal failure.
- neoplasia   • intestinal rupture.
- road traffic accidents.

Diagnosis

- transabdominal ultrasonography:
  - presence of excessive free abdominal fluid +/– a bladder wall defect (Fig. 3.35).
- abdominocentesis reveals peritoneal fluid:
  - usually containing with a low cell count and may smell like urine.
  - peritoneal creatinine concentration at least twice that of concurrently obtained serum sample is diagnostic for uroperitoneum.
  - infusion of new methylene blue into the bladder, followed by abdominocentesis 5–10 minutes later:
    - peritoneal fluid is blue-tinged, bladder or urachal rupture is confirmed.

- abdominal radiography only indicates free fluid in the abdomen (Fig. 3.36):
  - contrast cystogram in foals can confirm the site of urinary tract rupture.
- other diagnostic procedures to detect concomitant diseases, such as sepsis, bacterial peritonitis and urinary calculi.
- CBC results are usually normal if concurrent disease is not present.
- serum biochemical abnormalities usually include azotaemia, hyperkalaemia, hyponatraemia, hypochloraemia and metabolic acidosis.
- passive transfer of maternal antibodies should be evaluated in foals.
- umbilicus should be examined ultrasonographically.

Management

- initial treatment should be directed at stabilising the patient:
  - hydration should be maintained.
  - acid–base and electrolyte abnormalities should be corrected with i/v fluid therapy:
    - 0.9% or 0.45% saline should be used.
  - hyperkalaemic animals – dextrose-containing fluids are indicated (4–8 mg/kg/day):
    - severe hyperkalaemia (K^+ >5.5 mmol/l):
    - concurrent insulin (0.1–0.2 U/kg s/c) or sodium bicarbonate (1–2 mEq/kg).

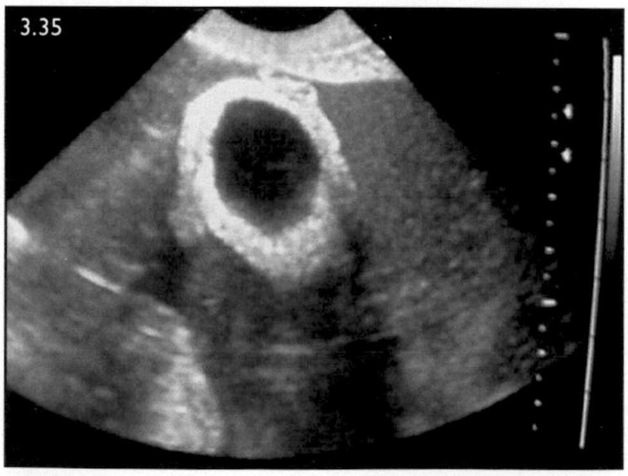

FIG. 3.35 Transabdominal ultrasonogram of a foal with uroperitoneum. The echogenic circular structure is the bladder surrounded by a large amount of free fluid in the peritoneal cavity.

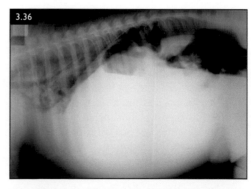

FIG. 3.36 Abdominal radiograph of a foal with uroperitoneum. The fluid line indicates the presence of free fluid in the abdomen.

- abdominal drainage is required in most cases (Fig. 3.37):
  - catheter placed in the abdomen and left in place until the defect is corrected.
  - abdominal drainage should be performed gradually:
    - intravenous fluid therapy should match amount of fluid removed from abdomen.
    - prevents acute hypotension following expansion of previously collapsed capillary beds.
    - peritoneal lavage helpful although catheter can be readily blocked by omentum and fibrin deposits.
- broad-spectrum antimicrobial therapy is indicated:
  - nephrotoxic drugs such as aminoglycosides should be avoided initially in azotaemic, hypotensive and dehydrated animals.
  - Ceftiofur sodium (2.2 mg/kg i/m or i/v q12 h) reasonable first-choice antimicrobial.
- surgical repair of the bladder defect should be performed after the animal's metabolic status has been corrected (Fig. 3.38):
  - in foals, internal umbilical remnants often a source of infection and should be removed during surgery.
  - laparoscopic repair of a bladder is less invasive and decreases time to recovery (Fig. 3.39).

- uroperitoneum should be managed similarly in adult animals:
  - laparoscopic procedures in standing horses have reported good outcomes.
- significant urine leakage from the ureters or kidneys is difficult to manage in horses:
  - if the defect persists, nephrectomy should be considered.

Prognosis

- good for recovery in foals.
- concomitant infection or sepsis significantly decreases a favourable outcome.
- in adults, or in cases of urine leakage from the kidney or ureters, less favourable.

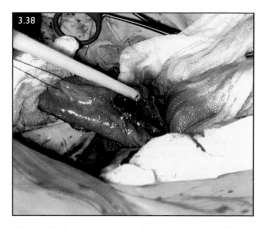

FIG. 3.38 Surgical repair of the bladder wall tear via laparotomy.

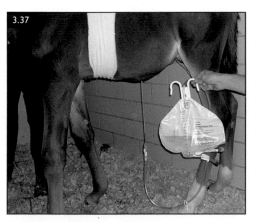

FIG. 3.37 Drainage of urine from the abdomen in a foal with uroperitoneum.

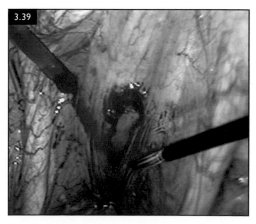

FIG. 3.39 Laparoscopic repair of a bladder wall tear in a foal.

Cantharidin toxicosis (blister beetle toxicosis) (see page 181, Book 3)

Definition/overview

- Cantharidin is a highly irritable substance that causes acantholysis and vesicle formation when in contact with skin or mucous membranes.
- compound contained in beetles belonging to the family Meloidae:
  - worldwide but toxicosis only a common event in the USA.
  - toxicosis when horses ingest food contaminated with blister beetles (*Epicauta* spp.), usually alfalfa hay.
- leads to acute renal tubular necrosis, renal injury and haematuria in the urinary tract.
- clinical signs:
  - shock.
  - GI and urinary tract irritation:
    - stranguria, pollakiuria and haematuria.
  - renal insufficiency, myocardial failure and hypocalcaemia.
  - sudden death with few prodromal signs may occur in some cases.
- multiple differential diagnoses:
  - cystitis   ○ pyelonephritis.
  - urolithiasis   ○ renal failure.
  - Neoplasia   ○ bladder paralysis.
  - NSAID and mercury toxicosis.
- diagnosis:
  - concurrent signs of urinary tract and GI disease.
  - particularly severe mucosal irritation.
  - examine hay for presence of blister beetles.
  - urinalysis:
    - urine SG is low (isosthenuria) even in the face of dehydration.
    - macroscopic haematuria usually present.

- epithelial cells occasionally seen but casts are rare.
  - hypocalcaemia, hypomagnesaemia, hypoproteinaemia and mild azotaemia.
  - chemical analysis for cantharidin by evaluation of urine or stomach contents:
    - submit samples quickly as cantharidin eliminated within 3–4 days.
- Management:
  - supportive treatment.
  - remove the source of toxin.
  - treat potentially exposed animals orally with mineral oil (4–6 litres via nasogastric tube) or activated charcoal (1–3 g/kg via nasogastric tube) but not concurrently.
  - intravenous balanced electrolyte solution administration (120–180 ml/kg/day).
  - administration of diuretics (furosemide 1 mg/kg i/v or i/m q6 h) to increase cantharidin excretion after patient is rehydrated.
  - supplementation of i/v fluids with calcium borogluconate may be required:
    - based on repeated evaluation of serum ionised calcium level.
  - magnesium supplementation is less commonly required:
    - administration of magnesium sulphate (0.2–1.0 g/kg dissolved in 4 litres of warm water q12 h).
  - analgesics administrated sparingly:
    - NSAIDs should be avoided if possible or given at reduced doses.
    - Alpha-2 agonists and opioids are good alternatives.
  - prophylactic treatment with appropriate antibiotics is controversial.
- variable prognosis and depends on:
  - amount of cantharidin ingested.
  - time from the onset of signs to the start of appropriate treatment.

Skin

INTRODUCTION

- usually, the owner will focus on and report one of the clinical syndromes shown below:
  - **Pruritus:** the 'itchy horse'.
  - **Pain:**
    - skin is painful to the touch.
    - variable degrees of pain associated with different types of disease/pathology.
  - **Hair density changes:**
    - excessive hairiness (the 'hairy horse').
    - loss of hair density and quality:
      - either generally or focally (alopecia or the 'balding horse').
  - **Pigmentary changes:**
    - either the hair or skin or both ('spotty horse').
  - **Scaling and flaking:**
    - dry dermatosis.
    - 'dandruff or scaly horse'.
  - **Moist dermatoses and crusting:**
    - weeping and seeping or moist dermatitis.
    - the 'eczematous horse'.
  - **Skin nodules:**
    - the 'lumpy or spotty horse'.
- owner's complaint is very significant for the clinical investigation, but it can be misleading as the owner may not have recognised every aspect of the case.
- skin conditions may be restricted to the skin or even a limited region of the skin:
  - **primary cutaneous disorders.**
- some skin disorders have a wider pathological involvement:
  - **skin disorders with a systemic implication.**
  - **systemic diseases with cutaneous manifestations.**
- important to establish which of these three categories of disease is present:
  - many cases may have subtle 'secondary' consequences.
- important genetic and hereditary skin diseases with significant clinical implications:
  - usually present at birth, or shortly afterwards (congenital disease), or develop later (developmental disease).
  - usually have a genetic origin (whether heritable or not) and may be difficult to manage the case and owner.
- **process of investigation of any skin condition must be logical and thorough:**
  - any system developed must ensure nothing is omitted and should be followed on every occasion.
  - useful to use a pre-prepared 'pro forma' sheet:
    - docs need to be comprehensive.
    - does not take a long time to complete if the clinician is logical and observant.
    - provides documentary evidence.
    - simple outline diagram of a horse from different angles provides an opportunity to locate and describe any lesions present.
    - photographs can easily be incorporated into digital clinical case records.
  - clinician must apply a logical interpretation of all of the signs that are detected and establish a **problem list.**
  - from this it is important to draw up a **differential diagnosis list:**
    - list should always be exhaustive even though some of the possibilities may be rare.
- fortunately, there are many economically accessible diagnostic aids in equine

DOI: 10.1201/9781003453666-4

dermatology that are easy to perform with minimal equipment:
- often help to achieve a definitive diagnosis.
- appropriate selection of these tests is important.
- concept of 'symptomatic treatment' must be avoided wherever possible.

Reaching a diagnosis

- only when a diagnosis is established can evidence-based information be provided for treatment and prognosis.
- awareness of common disorders encountered in any geographical or management environment is helpful.
- important throughout the process to keep an open mind.
- **Note it is important to recognise and accurately describe any lesion that is present, even if a definitive diagnosis cannot be attained.**

Primary lesions

- **Macule:** small (<0.5 cm diameter) circumscribed, flat, impalpable area of colour change in the skin (hypo- or hyper-pigmentation) (Fig. 4.1).
- **Patch:** similar to macule but >1 cm diameter with always some pigmentary change (Fig. 4.2).
- **Papule:** small (<1 cm diameter), circumscribed, solid, slightly raised mass in the skin (see Fig. 4.26):
  - viral papilloma or skin tag.
  - sometimes associated with insect bites.
- **Nodule:** raised, round solid lump >1 cm diameter (Fig. 4.3):
  - melanoma, sarcoid and neurofibroma.
  - eosinophilic granuloma and some allergic responses.
- **Plaque:** solid, elevated, flat-topped, regular or irregular thickening >1 cm diameter (Fig. 4.4):
  - usually allergic in origin, e.g. urticarial lesions.
- **Tumour:** mass of neoplastic origin (benign or malignant) (Fig. 4.5):
  - e.g. carcinoma, sarcoid, melanoma, papilloma.
  - usually not used for inflammatory or reactive masses.

- **Vesicle/bulla:** sharply demarcated, fluid-filled, raised blister-like lesion (Fig. 4.6):
  - 'bulla' if very large.
  - seldom seen because of their fragility.
- **Pustule:** vesicle filled with purulent material (or sometimes other cellular debris):
  - usually associated with bacterial infection (Fig. 4.7).
  - large accumulation of pus termed an abscess.
- **Wheal:** well-defined circular flat-topped oedematous lesions of variable size and shape (e.g. urticarial reactions) (Fig. 4.8).
- **Cyst:** closed sac-like or capsule structures that may be filled with semi-solid material or liquid:
  - usually an overlying pore (Fig. 4.9).

Secondary lesions

- **Scale:** accumulation of loose epidermis (Figs. 4.10, 4.11):
  - white or grey but sometimes discoloured by sebum, serum or blood.
  - usually result of imperfect cornification.
- **Crust:** solid, dry, adherent accumulation of dried serum, blood, pus or scales:
  - usually associated with skin injury (Figs. 4.12, 4.13).
- **Lichenification:** thickened and rough skin with prominent markings:
  - usually the result of repeated rubbing (Fig. 4.14).
- **Wound:** damaged skin arising usually from trauma (graze, bruise, laceration) and burns (see Chapter 5):
  - vary widely in size, severity and implications.
- **Abscess:** collection of pus that has built up within the tissue of the body (Fig. 4.15):
  - usually infection, but can be sterile.

Stage 1: Owner's complaints or concerns

- establish complaint or concerns:
  - usually, one or more of the recognised disease states mentioned above.
  - focuses the investigation process.
  - recorded but be careful it does not distort the approach to the case.

4

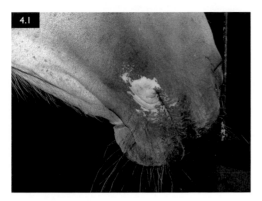

FIG. 4.1 Typical macular change in the lower right side of the face adjacent to the mouth.

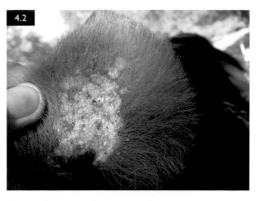

FIG. 4.2 Patch of depigmentation on the inner surface of the pinna.

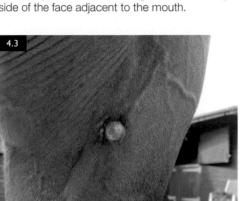

FIG. 4.3 Nodular sarcoid lesion on the inside of the upper forelimb.

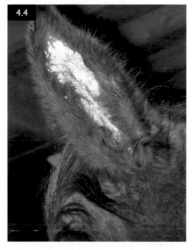

FIG. 4.4 An aural plaque lesion on the inner surface of the pinna.

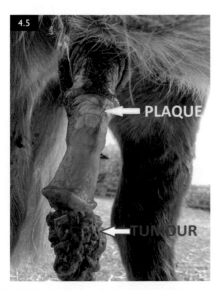

FIG. 4.5 Squamous cell carcinoma lesion of the glans penis with plaque lesions more proximally.

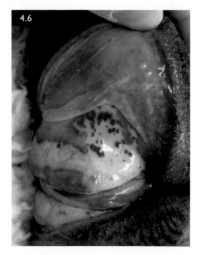

FIG. 4.6 A large collapsed blister or bulla lesion involving the oral mucosa of the upper lip in a young foal.

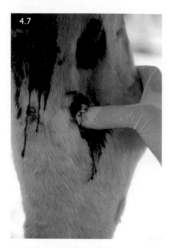

FIG. 4.7 Expression of pus from a pustule on the inner hock region following a stab incision. Note similar areas previously expressed more proximally.

FIG. 4.8 Multiple wheals on the right side of the shoulder and chest following an urticarial reaction.

FIG. 4.9 Small cystic structure on the lateral aspect of the hindlimb.

FIG. 4.10 Evidence of scaling on the skin and caught within the hair coat.

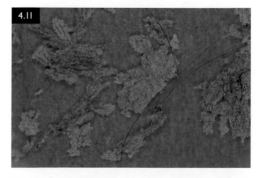

FIG. 4.11 Magnified view of skin scales of typical white or grey appearance.

FIG. 4.12 Severe crusting on a skin lesion involving the distal white-skinned hindlimb.

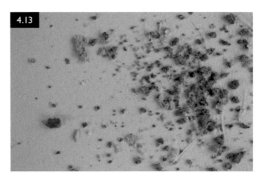

FIG. 4.13 Magnified appearance of the crusting after removal.

FIG. 4.15 Drainage of pus from an abscess on the tail-head after a stab incision.

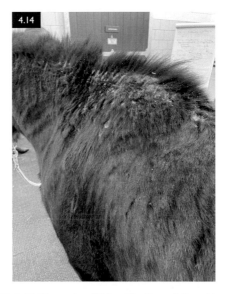

FIG. 4.14 Example of lichenification of the skin of the mane and withers region after repeated rubbing due to sweet itch.

Stage 2: Signalment (description) of the horse

- age • sex • breed • colour.
- approximate height and weight.
- focuses attention on diseases and complaints more likely in certain types and groups.

Stage 3: History

- thorough general clinical history and specifically for the skin condition.
- pre-prepared comprehensive examination sheet encourages thorough approach:
  - written or digital form provides definitive proof of the information supplied.
  - protects the clinician from 'deliberate or inadvertent' misleading information provided by the owner.

General history

- background to the horse from the time of ownership relating mainly to:
  - management (feeding, work/use and turn-out, etc.).
  - routine procedures (vaccinations, worming, dental or farriery).
  - travel and peer group health status:
    - may suggest common management practices, infections or toxicities.
  - current and recent health status of the horse:
    - any current medications or supplements.

Specific history of the primary complaint/complaints

- manner and timing of onset of the clinical problem and rate of progression.
- any treatment administered by the owner and response:
  - **owner may be reluctant to admit to anything.**

Stage 4: Clinical examination

- complete but basic clinical examination performed:
  - general health status of the horse must be established.
  - few minutes observing the horse from a distance to detect any behavioural signs.
- specific dermatological examination:
  - close examination of the skin in areas of normality and pathology.
  - establish the basic description, extent and number of skin abnormalities:
    - simple magnifying glass or the head of an auriscope:
      - focused lighting and magnification.
    - smartphone cameras and magnifying apps also useful.
  - lesions can be:
    - localised or generalised.
    - single or multiple.
    - isolated or coalescent.
    - not all lesions may be similar/identical:
      - co-morbidity can occur leading to a complex diagnosis.
  - lesions marked with their pathological description on an outline map of the horse.

Stage 5: The problem list

- all problems should be identified and recorded.
- irrelevant findings can be excluded, and the remaining problems assessed.
- remember individual 'problems' can form part of many different disease entities.

Stage 6: Differential diagnosis list

- list is formed from a series of possible conditions, no matter how rare.
- likelihood of the various options on the list should then be established:
  - some may then be reasonably eliminated.
- **Infectious disease:**
  - Viral.
  - Bacterial.
  - Fungal.
  - Protozoal.
  - Parasitic.
- **Non-infectious disease:**
  - Genetic and developmental conditions.
  - Immunological/allergic disease.
  - Traumatic conditions.
  - Nutritional/toxic or chemical disorders (deficiencies and excesses).
  - Endocrinologic disorders.
  - (Cardio)vascular disorders.
  - Neoplastic conditions.
- definitive diagnosis may be possible at this stage:
  - if not or diagnosis needs to be confirmed:
    - further diagnostic tests can be employed.

Stage 7: Selection of diagnostic tests

- various samples can be collected and examined.
- all add to the cost of the investigation and therefore must be focused and justified:
  - **does this test provide information needed to confirm or eliminate any differential diagnoses?**
  - **does it alter how the case is managed?**
  - **test not used if negative answer to either of the above.**

Sampling for parasitic disease

- all horses with pruritic skin disease or evidence of a papular or crusting dermatitis warrant investigation for parasitic skin disease.
- parasites sometimes directly visible, e.g. lice, ticks, fly maggots:
  - some require magnification, e.g. mites.
- use brushings, scrapings or tape preparations.
- samples taken from multiple fresh areas.
- negative findings may not rule out the presence of ectoparasites.
- severity of pruritus is not an indication of the parasitic load.
- some parasites are incidental/opportunistic.

Brushings

- harvest loose material and parasites from coat and skin surface using a stiff fine bristle brush, e.g. denture toothbrush or fine comb:
  - collected into container, tray or petri dish and then covered or sealed.
  - small handheld vacuum cleaner with a filter in the suction pipe is good for harvesting parasites.
- useful for identification of surface and blood-feeding lice:
  - does not detect parasites or fungal species within the dermis or epidermis.
  - examine promptly as the parasites can die quickly.

Brushings	
Advantages	• easy • limited equipment required • effective for large areas
Disadvantage	• requires prompt examination

Superficial scrapings

- collect hair, skin scurf and superficial layer of skin.
- surgical blade used to scrape epidermal material into a drop of mineral oil (liquid paraffin):
  - harvests a small amount of surface material but does not cause bleeding.
  - may detect lice and mites and useful for fungal infection.

Superficial scrapings	
Advantages	• easy • limited equipment required
Disadvantage	• limited harvest area • very limited value

Deep skin scrapings

- collect material from the intrafollicular space and superficial dermis:
  - collection of macerated accumulation of epidermal and dermal cells.
  - deep enough for bleeding to occur.
  - harvest a small area and are of limited diagnostic value.
- *Demodex, Sarcoptes, Chorioptes* spp., *Pelodera* spp. or *Strongyloides* spp. may be found in these samples as they are often restricted to the dermis.

Deep scrapings	
Advantages	• easy • limited equipment required • **only** way of sampling for burrowing parasites apart from biopsy
Disadvantage	• small area of harvest • sometimes painful

Clear adhesive acetate tape preparation

- mainly used to detect *Oxyuris equi* eggs in the perineum.
- useful for other free-moving skin parasites:
  - Lice and *Dermanyssus* spp. (poultry red mites), *Chorioptes* spp. *Psoroptes equi,* and *Trombiculid* mites
- after removal from skin, the tape is placed on a microscope slide with a drop of mineral oil (liquid paraffin):
  - examined also by a hand lens or handheld device app.

Sticky-tape testing	
Advantages	• easy • limited equipment required • effective sampling from wide area
Disadvantage	• surface only • requires prompt examination • often messy • limited efficiency • tedious microscopy unless dissecting microscope is available

Hair and crust collection

- plucks of hair and collection of crusts:
  - indicated where broken hairs, and crust and/or scale are present.
- preferred samples for detection of:
  - dermatophytes *Trichophyton* spp. and *Microsporum* spp.
  - *Dermatophilus* spp. and other bacteria.
- hair and associated crusts plucked with sterile haemostats from a number of fresh lesions, preferably from the margins.
- placed in sterile containers:
  - unsealed for suspected fungal investigations.
  - sealed in microaerophyllic state for *Dermatophillus* spp.
- examined by direct microscopy:
  - appropriate staining to allow identification of gross/microscopic organisms.
  - specific stains for bacterial and fungal elements to improve diagnostic accuracy.
- **Hair samples:**
  - cultured for bacterial and fungal species:
    - ◆ standard agar culture for bacterial organisms in appropriate conditions:
      - *Dermatophilus* spp. microaerophilic conditions.
      - anaerobes require anaerobic conditions.
    - ◆ Dermatophyte culture requires Sabouraud's dextrose agar:
      - up to 14–28 days to establish a diagnostic colony.
      - media with phenol red pH indicator available commercially:
        - ◇ may achieve an earlier result.
  - direct microscopy also used with/without clearing the sample with potassium hydroxide (KOH).

- real-time polymerase chain reaction (qPCR) to detect presence of dermatophytes:
  - ◆ accurate and efficient.
  - ◆ may allow rapid speciation of positive samples.

Hair and crust sampling	
Advantages	• easy • limited equipment required • effective sampling from large area
Disadvantage	• requires microscopic or culture methodology

Swabs for bacterial and fungal culture

- skin normally not sterile and therefore bacteriology requires careful interpretation.
- normal microbiome:
  - commensal organisms essential for skin health.
  - opportunistic pathogens.
- primary pathogens are uncommon.
- **single species cultures often pathological.**
- **complex cultures often less significant.**
- suitable samples include:
  - skin biopsies (avoid sterilising the sample before collection).
  - needle aspirates of pustules.
  - swabs from freshly ruptured pustule or beneath a freshly lifted crust.
- culture and sensitivity patterns do not always correlate well with efficacy.
- *Dermatophilus congolensis* culture requires microaerophilic conditions:
  - sample kept in a sealed container with low oxygen concentrations.
- fungal cultures are more problematic.

Culture methods	
Advantages	• easy • limited equipment required • effective sampling from focused areas • multiple samples are easily collected
Disadvantage	• requires laboratory examination even if direct smears are taken

Cytology

- smears from erosions or ulcerated areas or draining/exudative lesions.
- directly onto a slide from the lesion or its cut surface.
- Fine-needle aspirates from the contents of a nodule or mass:
  ○ 2 ml syringe and a 22–23 gauge needle.
  ○ not generally helpful for solid masses:
    ♦ better to introduce the needle a few times into the lesion.
    ♦ then quickly 'blast' needle contents onto a pre-marked and identified slide.
    ♦ **remember to mark frosted end of slide with a pencil NOT marker pen.**
  ○ aspiration is carried out and blood obtained – sample probably useless.
- often unrewarding especially with poor technique.
- weak diagnostic aid for horses unless a skilled cytopathologist is available:
  ○ tissue biopsy may be a more practical and reliable technique.

Cytology	
Advantages	• easy • limited equipment required
Disadvantage	• easy to do badly • poor technique leads to very poor outcome • limited value in horses

Biopsy for histopathology

- useful for the diagnosis of:
  ○ nodular and diffuse crusting lesions.
  ○ areas of inflammatory reactions of all types.
- not useful for:
  ○ chronic lesions.
  ○ those with extensive secondary changes such as self-inflicted trauma.
- results depend on the care taken in collection:
  ○ appropriate technique from representative areas of pathology.
- different biopsy specimens:
  ○ **Excisional** where whole of the lesion is removed.
  ○ **Portional** where a portion of the pathology is removed:

- ♦ can include a margin of the pathology if considered appropriate.
  ○ **Shave** – shave layers parallel to the skin surface until appropriate depth reached:
    ♦ inform pathologist.
    ♦ useful for coronary band pathology.
  ○ **Hollow needle** (Tru-Cut) biopsy:
    ♦ dedicated manual, spring loaded or hydraulic system of an appropriate size (usually 12–16 g).
    ♦ useful for solid masses where the skin itself is normal or not wise to remove any overlying skin.
- multiple samples help eliminate potential errors of collection:
  ○ minimum of 4–6 samples usually submitted.
- biopsy punches are often used for excisional or portional biopsy:
  ○ size of instrument dictated by type/location of lesion:
    ♦ 4–6 mm for distal limb and eyelid.
    ♦ 8 mm elsewhere.
    ♦ transmission risk if same instrument used for multiple biopsies:
      – new instrument for each site (e.g. sarcoid).
  ○ pathology and type of disease:
    ♦ 4–6 mm on the distal limb and the eyelid regions.
    ♦ 8 mm for general use elsewhere.
    ♦ multiple biopsies risk transmission of the condition from one site to another:
      – some cases (e.g. sarcoid) require new instrument for each sample.
- **always submit all clinical data and if possible clinical images with the biopsy:**
  ○ contact the pathologist for advice pre biopsy regarding sampling:
    ♦ different tissue fixatives and despatch instructions may be required.

Technique

- do not clip, wash or wipe the site as may remove significant diagnostic material:
  ○ crusting is present:
    ♦ important not removed by skin preparation and cleaning.

- ♦ key diagnostic features may exist in the most superficial layers.
- sedation and local anaesthetic (LA) may be required:
  - ○ small volumes of LA placed locally and carefully.
  - ○ needle left in place to identify the site.
  - ○ sampling carried out just distal to needle point.
- scalpel or circular biopsy punch (4-6-8-11 mm diameter) used to sample the skin:
  - ○ fine non-toothed forceps or a needle used to carefully remove the biopsy.
- samples for histopathology usually placed in 10% buffered neutral formal saline.

Tissue biopsy	
Advantages	• easy and safe • limited equipment required • definitive pathological description of the skin changes
Disadvantage	• may **not** lead to a definitive diagnosis • less useful for chronic lesions and those with secondary changes such as self-inflicted trauma and secondary infection

- **Immunohistochemistry (IHC) staining** of tissue samples:
  - ○ combines anatomical, immunological and biochemical techniques.
  - ○ identification of specific cellular types and responses in tissues:
    - ♦ appropriately labelled antibodies and stains bind specifically to target antigens *in situ*.
  - ○ can be coupled with direct or indirect fluorescent antibody methods in autoimmune- or immune-mediated conditions such as pemphigus.
  - ○ often accomplished on standard formalin-fixed tissue samples.
- **PCR:**
  - ○ detects specific DNA/RNA extracted from a sample, including formalin-fixed biopsies.
  - ○ can detect viral, rickettsial, bacterial, parasitic, protozoal or fungal organisms.

Antinuclear antibody testing (ANA)

- ANA titre and lupus erythematosus (LE) cell detection assist in the diagnosis of some autoimmune disorders, e.g. lupus syndromes:
  - ○ ANA titre detects a component of the cell nucleus.
  - ○ LE preparation detects antibody to nucleoprotein.
  - ○ positive (titres >1:160) indicates possible autoimmune dysfunction.

Allergy testing
Intradermal skin testing in horses

- tests local immune response in the skin.
- currently the only reliable method of identifying allergens responsible for disease.
- may lead to preparation of an immunotherapy treatment extract.
- small amounts of various diluted relevant allergens are injected intradermally:
  - ○ compared to positive and negative controls.
  - ○ commonly read at 1, 6 and 24 hours.
  - ○ most reactions occur at:
    - ♦ 30–60 minutes (immediate responses).
    - ♦ and/or 4–6 hours (late responses).
- commonly, false-negative, false-positive or subclinical positive tests occur.
- results always interpreted alongside the clinical history.
- horse must not have been recently treated with drugs that might influence the results:
  - ○ glucocorticoids (4 weeks).
  - ○ antihistamines and some tranquillisers (2 weeks).

Immunoglobulin E (IgE) blood sampling

- all allergy tests performed on blood samples have significant limitations.
- several immunological 'steps' away from site of disease and may be misleading.
- many tests have very unreliable specificity, sensitivity and repeatability.
- results can vary widely between and within individual laboratories.

GENETIC/HERITABLE SKIN DISORDERS

Congenital papillomatosis

Definition/overview

- single or multiple, papilloma-like lesions on the skin present at birth.
- variable size.
- may be found on the placenta.

Aetiology/pathophysiology

- congenital condition.
- controversial aetiology as no active virus is found in these lesions.

Clinical presentation

- usually, single solitary papilloma-like lesions present from birth on head or body.
- little/very slow progression.

Diagnosis

- location and clinical picture are suggestive (Fig. 4.16).
- excisional biopsy is definitive.
- can be confused with cavernous haemangioma.

Management

- early surgical excision is required.
- no recurrence if surgery performed correctly.

Prognosis

- very good following surgery.
- benign neglect not recommended.

Cutaneous agenesis/ epitheliogenesis imperfecta

Definition/overview

- sporadic and very rare.
- defects within embryological skin formation.

Aetiology/pathophysiology

- genetic abnormality not known to be familial.

Clinical presentation

- at birth, absence of skin over variable areas – most often one or more limbs (Fig. 4.17).
- little/no inflammation but often rapidly infected by opportunistic pathogens and result in septicaemia.

Diagnosis

- characteristic appearance often mistaken for birth/handling trauma.

Management

- no treatment is effective – attempts to skin graft the sites are hopeless.

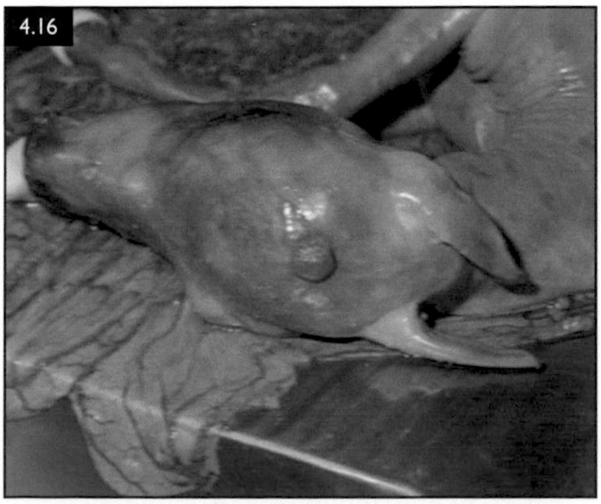

FIG. 4.16 Papillomatosis. A 10 mm wart on an aborted foal's head is shown.

FIG. 4.17 Aplasia cutis congenita (epitheliogenesis imperfecta). Complete absence of epidermis and skin appendages distal to the carpus and tarsus. (Photo courtesy JP Hughes.)

Prognosis

- hopeless.
- secondary infection can result in sepsis and septicaemia.

Epidermolysis bullosa

Definition/overview

- group of heritable disorders resulting in mechano-bullous epidermal changes at, or shortly after, birth.
- autosomal recessive gene is suspected.
- Belgian and American Saddlebred horses.
- sporadic cases in other breeds.

Aetiology/pathophysiology

- pathological changes are divided into 3 basic patterns:
  ○ classified by the location of the epidermal cleavage and blister formation.
- histological diagnosis is challenging.

Clinical presentation

- wide variation in extent and severity (Fig. 4.18).
- foals are born with epidermal blisters that rupture easily leaving well demarcated, open ulcerated area of skin and mucosal collarettes.
- not always obvious at birth.

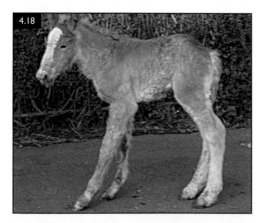

FIG. 4.18 Junctional epidermolysis bullosa occurs in Belgian draught foals, with lesions commonly occurring at the skin and mucocutaneous junctions. (Photo courtesy DC Knottenbelt; reprinted from Pascoe RR and Knottenbelt DC (1999) *Manual of Equine Dermatology*, WB Saunders, with permission.)

- limbs most often affected – particularly the coronary band tissues:
  ○ may lead to sloughing of the hoof capsules.
- oral lesions are often multiple with small areas of blister formation:
  ○ may rupture early with oral bleeding and blood in the saliva.
- dental and corneal changes may also be present.

Diagnosis

- changes present at birth or within first 4–6 weeks of life without healing highly suggestive.
- can be mistaken for juvenile (neonatal) pemphigus and neonatal neutropenic epidermal bullous disease:
  ○ former is histologically characteristic.
  ○ latter usually resolves very quickly.
  ○ **important diagnosis is correct as outlook for each is very different.**

Management

- affected areas kept covered to avoid secondary infection whilst awaiting diagnosis.
- disease is regarded as incurable:
  ○ sites do not heal and become progressively more extensive.
  ○ euthanasia is indicated in confirmed cases.
- breeds where the disease is recognised:
  ○ **stallion should not be used for breeding.**
- genetic test is available to detect carrier status animals.

Prognosis

- disease is heritable and carries a hopeless prognosis.

Hereditary equine regional dermal asthenia (HERDA) (also known as cutaneous asthenia and hyperelastosis cutis)

Definition/overview

- group of autosomal recessive inherited connective tissue disorders.
- Quarter Horse breed most affected:

4

- ○ Paints and Appaloosas with Quarter Horse lineage also reported.
- other breeds reported sporadically include:
  - ○ Arabian and crosses.
  - ○ Thoroughbred and crosses.
  - ○ Hanoverian and Haflinger.

Aetiology/pathophysiology

- autosomal recessive mutation inheritance:
  - ○ resulting in a functional protein defect.
  - ○ disturbs structural organisation and orientation of collagen fibres:
    - ♦ with/without decreased repair mechanisms resulting in poor/non-healing sites.
- affected horses have lower cutaneous tensile strength than healthy horses.

Clinical presentation

- signs appear from birth or may be delayed until horse is ridden for the first time with harness/tack.
- often present with loose, wrinkled skin, which is hyperextensible, hyper-fragile:
  - ○ tears easily and repairs very slowly (if at all).
- older horses usually present with:
  - ○ inexplicable wounds.
  - ○ scarring in friction areas under the saddle, over the back and sides of neck.
- may be generalised over the whole body or sharply demarcated in limited regions.
- abnormal (non-contracting) scar formation is common.
- single or multiple, subcutaneous haematomas and abscesses may occur (Fig 4.19).
- may result in defects in the cornea, heart, tendons and great vessels.

Diagnosis

- breed specific skin changes are characteristic and well recognised by breeders.
- loose, hyper-fragile skin is suggestive.
- difficult to distinguish affected from healthy horses on histopathology alone.
- genetic testing available to determine carrier or affected status at the Veterinary Genetics Laboratory at UC Davis (USA):
  - ○ 20–30 hairs with roots should be submitted.

Management

- no treatment is effective apart from minimising trauma and palliative wound site management to limit secondary infection.
- attempts to graft affected areas or otherwise encourage healing are fruitless.
- remove affected horses from breeding programmes.

Prognosis

- hopeless.
- all affected horses should be destroyed even if the regions affected are limited.

Lethal white syndrome

Definition/overview

- fatal heritable disorder affecting Overo-Paint horses.
- also known as ileocolonic aganglionosis often with concurrent atresia coli (see page 183, Book 3).

Aetiology/pathophysiology

- neonatal foal of Overo–Overo American Paint horse breeding with double recessive genes.
- complete loss of function of the endothelin receptor type B (EDNRB) gene.

Clinical presentation

- foals are born completely white with blue eyes or sometimes with faint brown marking on the ears or elsewhere.

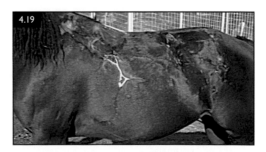

FIG. 4.19 Cutaneous asthenia in a 3-year-old inbred Quarter horse filly (dam × own son). There is abscessation of loosened skin and the presence of old scars from previous similar events. (Reprinted from Pascoe RR and Knottenbelt DC (1999) *Manual of Equine Dermatology*, WB Saunders, with permission.)

- often normal at birth but rapidly develop severe progressive and non-responsive colic.

Diagnosis

- characteristic coat colour and history.
- differential diagnosis mainly related to either meconium impaction, intestinal atresia or other causes of severe colic and bloating within hours of birth.

Management

- no treatment is available.
- euthanasia should be performed immediately.

Prognosis

- 100% fatal disorder.
- genetic tests are available to detect carrier horses which should not be used for breeding.

Dermoid cysts

Definition/overview

- single or multiple, firm to fluctuant, smooth, round cysts, usually with normal overlying haired skin.
- most commonly along the dorsal midline of the thorax, lumbar and croup regions.
- breed predisposition not proven but may be more frequent in Thoroughbred yearlings.

Aetiology/pathophysiology

- congenital, developmental and possibly hereditary lesions.
- embryonal displacement of ectoderm into the subcutis.

Clinical presentation

- nodules, occasionally coalescent, 10–15 mm in diameter.
- usually found on the dorsal midline from the withers to rump.
- contain soft, keratinaceous, grey material and coiled hairs.
- most often detected between birth and 18 months.

Diagnosis

- location and clinical picture are characteristic.

- ultrasonography is useful.
- excisional biopsy shows a cyst wall lined with stratified squamous epithelium containing variable adnexal structures (Fig. 4.20).

Management

- complete surgical excision of the cyst:
  - recurrence should not occur if excision is complete.

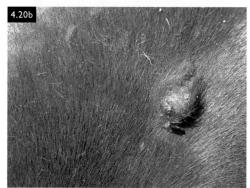

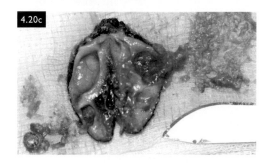

FIG. 4.20 (a) Dermoid cyst in the dorsal midline of a young horse. (b) Detailed view. (c) After removal and following transection. Note the contents with hair, sebum and keratin.

Prognosis

- very good following surgery.
- benign neglect is not recommended.

Epidermoid cysts

Definition/overview

- usually solitary cysts, 7–30 mm in diameter, occurring on the head and limbs.
- freely movable, well-circumscribed, and firm to fluctuant on palpation.
- often contain a yellow to grey mucoid fluid.
- Atheroma is a type of epidermoid cyst that occurs in the false nostril (see page 11, Book 3).

Aetiology/pathophysiology

- cause is unknown.
- classified according to differentiation of the epithelial lining.

Clinical presentation

- large nodules, usually single, and containing only mucoid fluid and no hairs.

Differential diagnosis

- other cysts: dentigerous, false nostril, dermoid or conchal.
- Hypodermiasis.

Diagnosis

- location and character are pathognomonic but can be confirmed by histopathology.

Management

- often incidental and can be left alone.
- very large cysts may be clinically significant and may require surgical removal.

Prognosis

- excellent providing all of the cystic lining material is removed.

Dentigerous cysts
(see page 12, Book 3)

Definition/overview

- firm or cystic swellings developing in the temporal region between the base of the ear and the eye arising from aberrant dental germ tissues.
- occasionally also found on the cranial vault or within the maxillary sinus.
- may contain obvious dental tissues or have a more cystic nature with characteristic draining sinus on the lower third of the leading edge of the pinna.
- surgical removal is usually curative, but healing and recurrence can be a problem.

Calcinosis circumscripta
(see page 190, Book 1)

Definition/overview

- rare cutaneous or subcutaneous solid plaques or localised swellings often over joints particularly the lateral stifle region (Fig. 4.21).
- usually in younger horses (1–5 years).
- all breeds may be affected.
- usually no lameness, but slowly progressive in some cases.
- characteristic radiographic and ultrasonographic features:
  - calcification with a well-defined border.
- differential diagnosis may include mast cell tumour and eosinophilic granuloma.
- often asymptomatic and can be left alone.
- surgical removal may be indicated but there can be significant complications.

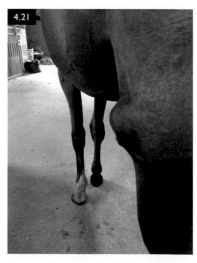

FIG. 4.21 A 4-year-old Anglo-Arab gelding which had a long history of a mass on the outside of the gaskin which was diagnosed as calcinosis cutis.

Linear keratosis/ Epidermal naevus

Definition/overview

- occasional disorders of keratinisation occurring in various forms including linear and wider, more focal, forms.
- linear keratosis is a linear area of hair loss and hyperkeratinisation (Fig. 4.22):
  - most often on shoulder, neck or buttock.
- relatively common form known as 'Mallenders and Sallenders' in heavy or cob breeds:
  - palmar aspect of the carpal regions (Mallenders).
  - dorsal aspect of the tarsus (Sallenders).
  - very rare in other types and breeds.

Aetiology/pathophysiology

- disorders of keratinisation.
- inherited genetic basis is suspected.

Clinical presentation

- linear or focal areas of hair loss with characteristic 'pouring' patterns over side of the neck, shoulder, quarter and occasionally other areas, including the face and limbs.
- often a single 'line' of the condition but occasionally broader and multiple:
  - wider more defined form seen on palmar and plantar aspect of cannon region.
- Mallenders and Sallenders presents with scaling and cracking of the skin behind the knee and in front of the hock.

Differential diagnosis

- mistaken for other scaling conditions including dermatophytosis, sarcoid and sarcoidosis.
- scale feeding mites such as *Chorioptes* spp. are often found at these sites but are not the cause of the problem:
  - may induce pruritus and self-trauma with continued deterioration of both primary and secondary signs.

Diagnosis

- characteristic appearance and history for all forms of the condition.

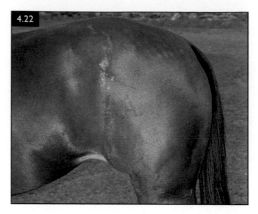

FIG. 4.22 Equine linear keratosis is seen as vertical linear bands of hyperkeratosis.

- characteristic histopathology from biopsy samples.

Management

- treatment can make the condition worse.
- skin should be managed gently with minimal trauma.
- Retinoids and urea-based creams and gels are sometimes used to improve the cosmetic appearance but often leave a deteriorating legacy.

Prognosis

- lifelong condition that does not usually warrant interference:
  - incidental cosmetic nuisance.
  - repeated attempts to 'cure' can lead to serious local scarring.

Follicular dysplasia (mane and tail dystrophy) (Fig. 4.23)

Definition/overview

- genetic, heritable disorder of hair follicle development leading to:
  - hair loss or poor-quality hair coat density.
- most often in Appaloosa horses.
- present at birth in some cases but may become more obvious later in life.
- affects mane and tail (Appaloosa Mane and Tail Dystrophy) or other regional or general sites (hypotrichosis).

4

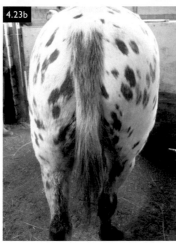

FIG. 4.23 Typical appearance of mane (a) and tail (b) dystrophy of an Appaloosa horse aged 8 years old.

Aetiology/pathophysiology
- genetic and heritable.
- usually remains static but occasionally it can be progressive.
- distinctive abnormal hair follicle structure, shape and size with dysplastic hair shafts.

Clinical presentation
- sparse hair density and poor-quality, short or brittle hairs.
- mane and tail often affected in Appaloosa horses and their crosses.

Differential diagnosis
- Alopecia areata.
- scarring and other traumatic or other insults that induce hair alteration and hair loss.
- some forms of sarcoid or pemphigus syndromes.

- some rare toxicities affect the mane and tail hair in particular:
  - selenium toxicosis.
  - arsenic and mercury poisoning.

Diagnosis
- characteristic appearance:
  - often changes are subtle and easily missed.
- biopsy is helpful.

Management
- no treatment.
- often little or no change over time.
- Mane and tail dystrophy of Appaloosa horses is usually slowly progressive.

Prognosis
- most horses are otherwise unaffected, and it is a cosmetic nuisance.
- Mane and tail dystrophy may be more profound and affect wide areas of the body.

Fading Appaloosa syndrome

Definition/overview
- heritable condition of Appaloosa horses in which progressive changes occur in hair colouration without any obvious changes in skin colour.

Aetiology/pathophysiology
- heritable genetic basis which is more common in some breed lines.

Clinical presentation
- progressive change in colouration and patterning occurring over some years.

Diagnosis
- typical for breed but changes may be subtle.
- photographic evidence of change highlights the changes (Fig. 4.24).

Management
- no treatment is effective or warranted.

Prognosis
- general health is unaffected.

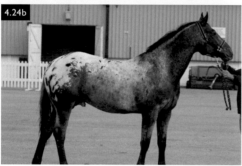

FIGS. 4.24 Typical changes that Appaloosa horses develop in their hair colour over time. Same horse taken as a foal (a) and 5 years later (b).

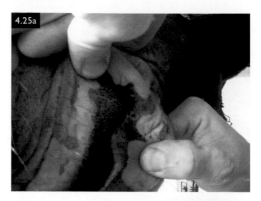

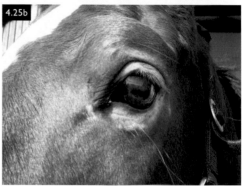

FIG. 4.25 Typical case of vitiligo affecting the side of the mouth (a) and around the eyes (b). Changes advanced slowly for 3–4 years and thereafter remained static.

Vitiligo/Leucoderma (and other idiopathic pigmentary changes)

Definition/overview
- idiopathic, asymptomatic, non-inflammatory change in skin colouration.
- Arabian horses are commonly affected (Fading Arab Syndrome) but other breeds are occasionally affected as well.

Aetiology/pathophysiology
- genetic basis but otherwise pathogenesis is unknown.

Clinical presentation
- small or larger areas of pigmentary changes most commonly around the eyes and face (mouth and cheeks) (Fig. 4.25).

Differential diagnosis
- autoimmune conditions.
- copper deficiency.
- difficult to differentiate from changes that follow historical trauma or surgery.
- acquired leucoderma follows freeze-branding and some other skin insults including localised chemotherapy.

Diagnosis
- changes are benign and with no history of trauma or other skin inflammatory condition.

Management
- no treatment for any change in skin or hair pigmentation.
- affected areas are best left alone.

- topical applications of bleomycin have been shown to result in skin darkening.
- cosmetic cases can be treated with suitable hair dyes used with caution.
- protection from sunlight exposure may be necessary.

Prognosis

- horses are unaffected by the condition of genuine idiopathic hair coat changes (Vitiligo).
- depigmented skin may be more liable to actinic dermatitis and carcinoma development.

COMMON VIRAL SKIN DISEASES

Papillomatosis

Definition/overview

- characterised by single or multiple, isolated or coalescent wart lesions of variable size.
- most often found on the nose, lips, around the eyes and inside the ear:
  - less frequently on the neck and limbs.

Aetiology/pathophysiology

- seven different *Equus caballus* papillomaviruses (EcPV) have been identified.
- three clinical presentations are recognised:
  - Classical warts (Grass warts).
  - Genital papillomata, plaques, carcinoma *in situ* and squamous cell carcinoma (SCC).
  - Aural plaques.
- transmission via:
  - direct contact with an infected horse.
  - indirectly through contaminated fomites.
  - sexual transmission.
  - trauma including insect bites.

Clinical presentation

- EcPV1 papillomatosis is characterised by small, cauliflower-like, warty growths that increase in number rather than size:
  - predominately in young horses (9–36 months of age) (Fig. 4.26).
  - occasionally affect mature to aged horses.
- EcPV2 and sometimes EcPV7 genital lesions are encountered:
  - multiple or isolated individual or confluent.
  - greyish papules or plaques of coalescent papillomata.

- genital lesions more frequent in the free portion of the genitalia in male horses (either geldings or stallions) and around the vulva in mares.
- Aural plaques associated with EcPV3 to EcPV6:
  - well-demarcated, erythematous or depigmented multiple or coalescent lesions:
    - affect inside surface of one or both pinnae (Fig. 4.27).
  - asymptomatic in most cases.
  - occasionally become sore and associated with behavioural problems including head shaking signs.
  - quite frequently found in the groin region of geldings:
    - very extensive on the sheath, the abdominal wall and the medial thighs.
    - unusual in mares.

Diagnosis

- clinical appearance, age of the horse, and location, size and number of wart-like lesions are characteristic.

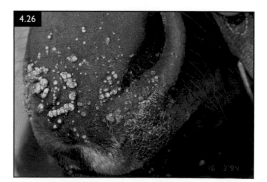

FIG. 4.26 Three-month-old lesions of papillomatosis around the upper and lower lips of a weanling Thoroughbred filly.

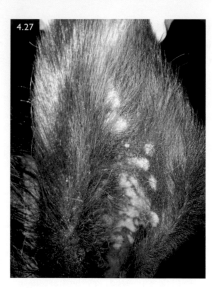

FIG. 4.27 Typical aural plaques within the skin of the inner pinna.

- histopathology of biopsy specimens required to rule out other disorders:
  - verrucous sarcoid.
  - *in situ* or invasive SCC in cases of genital papilloma/papillomatosis.
- PCR from a swab, skin scraping, or biopsy will confirm the diagnosis in most cases:
  - negative results can arise.

Management

- most EcPV1 lesions in young horses usually resolve spontaneously:
  - within 3–4 months unless patient significantly immunocompromised.
  - older horses less likely to undergo spontaneous (immunological) resolution.
  - surgical removal of warts from around the eyes and commissure of mouth occasionally required.
  - podophyllin, salicylic acid and imiquimod have been used in persistent lesions.
- fly control is important, especially when treating aural plaques.
- genital warts, and consequent *in situ* carcinomas, can be treated with imiquimod:
  - often extremely painful and should be used with caution.

- surgery recommended when SCC is diagnosed:
  - laser surgery and cryotherapy can be effective.
- other therapies with reported variable responses include:
  - bloodroot extract (an escharotic/ caustic agent).
  - intralesional cisplatin, IL-2 or *Propionibacterium acnes*.
  - topical retinoids such as tazarotene or retinoin.

Prognosis

- excellent in EcPV1 papillomatosis as most cases regress spontaneously.
- aural plaques generally do not spontaneously regress but are usually well-tolerated by the horse:
  - cosmetic issues can be significant.
- genital papillomata that transform into *in situ* or invasive SCC carry a more guarded prognosis.

Equine molluscum contagiosum (Uasin Gishu disease)

Definition/overview

- mildly contagious cutaneous infection caused by an unclassified poxvirus.
- less common in temperate climates but becoming more prevalent.

Aetiology/pathophysiology

- active viral infection restricted to the skin.

Clinical presentation (Fig. 4.28)

- widespread, multiple, isolated or coalescent, papular wart-like cutaneous lesions:
  - lesions often have a waxy appearance.
- progression in numbers and extent of lesions over some months.
- otherwise, asymptomatic and no systemic effects.

Differential diagnosis

- difficult to differentiate from viral papilloma, verrucose sarcoid and parapox virus infection.

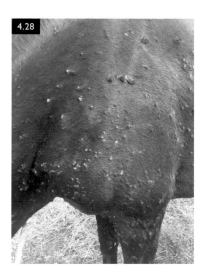

FIG. 4.28 One of several horses affected by molluscum contagiosum in a stable yard. The lesions resolved over 2–3 years but left some obvious wart-like lesions permanently.

Diagnosis

- geographically restricted but occasionally found outside those regions.
- histological diagnosis is characteristic with molluscum bodies in the epidermis.
- PCR methods on tissue samples or swabs can be definitive.

Management

- treatment is unrewarding and most cases are left to recover on their own.
- fly control (probably) important to control spread across the horse and horse to horse.
- topical imiquimod is possibly useful but impractical where there are widespread multiple small lesions.

Prognosis

- generally good but may take up to several years to resolve.
- minimal clinical effects on the horse even when severely affected.

Equine coital exanthema

(see pages 81–83, Book 2)

Definition

- venereal transmitted viral skin disease caused by equine herpesvirus (EHV)-3.

- EHV-3 can be transmitted during coitus, nasal contact or fomites (Figs. 4.29, 4.30).

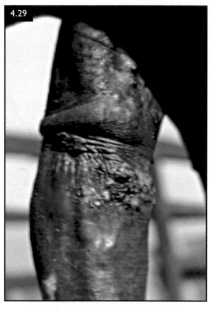

FIG. 4.29 Equine coital exanthema. Infection occurred in a 4-year-old first-season stallion approximately 12 days post-mating with an infected mare.

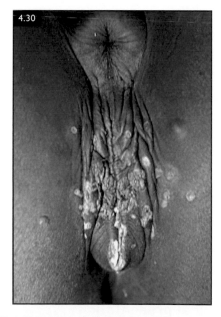

FIG. 4.30 Equine coital exanthema. Papules, pustules, crusts and erosion of the skin related to the vulva, anus and ventral surface of the tail have developed in this post-service.

- vesicles occur on the penis of stallions and the vulvar region of mares around 5–10 days after an infected mating:
  - lesions can also develop around the nose and mouth.
  - sometimes markedly painful and may discourage breeding.
  - rapid progression of vesicles to small focal necrotic areas.
  - severe cases can affect the anus and tail skin.
  - usually by day 4, the lesions are covered with serum and crusts.
  - granulation occurs within 10–14 days.
  - some lesions may leave a small focal depigmented scar.
- virus isolation only successful when samples are taken as soon as lesions appear:
  - PCR methods are helpful.
- paired serum samples 28 days apart for serum neutralisation and complement fixation testing:
  - increased titres indicative of recent infection.
- gently clean off crusts with accelerated hydrogen peroxide:
  - eroded areas may benefit from topical application of iodine-based solutions.
- topical daily antibiotic and corticosteroid ointment can be used to treat any inflammation and secondary bacterial infections.
- sexual rest until the lesions have healed, usually 14–21 days.
- prognosis is good with lifelong immunity in both stallions and mares.

COMMON BACTERIAL SKIN DISEASES

Staphylococcal skin infection

Definition/overview

- *Staphylococcus aureus, S. pseudintermedius, S. delphini* and *S. hyicus* are the common isolates.
- Methicillin-resistant *Staphylococcus aureus* (MRSA) may also be identified in natural and nosocomial infections.

Aetiology/pathophysiology

- usually associated with unhygienic conditions:
  - dirty harness, grooming equipment, rugs and blankets etc.
- may be secondary to underlying inflammatory and metabolic skin conditions (e.g. atopy, insect-bite hypersensitivity).
- bacteria gains entry via a contaminated wound, skin abrasion and other less serious skin disorders such as dermatophytosis.

Clinical presentation

- small (1–2 mm) lesions that rapidly enlarge with accompanying local oedema and significant pain:
  - local lymphatic vessels can become prominent.
- some lesions may exude plasma and large areas may coalesce (Fig. 4.31):
  - lesions are often associated with harness areas and saddle cloths ('saddle rash').
  - commonly affect the skin of the back, saddle area, loins and chest.

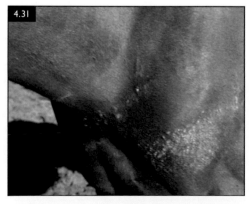

FIG. 4.31 Staphylococcal folliculitis. The shoulder of a draught horse with small alopecic pustules with a number of adjacent small papules due to infection following rug abrasion.

- crusting over the surface mats the hair together:
  - usually, little pus is present.
  - crusts can dry, become adherent and difficult to dislodge.
- *S. hyicus* has been isolated from 'greasy heel'-type lesions of the pastern and coronet:
  - may be pruritic and severely painful.

Differential diagnosis

- Streptococcal folliculitis.
- other bacterial and fungal infections.

Diagnosis

- pain is a significant feature of all staphylococcal infections of the skin.
- direct smears and cultures from swabs taken from the surface and deep in the lesions will reveal the organism responsible.

Management

- focal abscesses drained and thoroughly cleaned with antibacterial washes using ten standard hot wash system (see page 119).
- Potentiated sulphonamide is the most used first-line antibiotic but achieving effective minimum inhibitory concentration (MIC) levels is difficult:
  - 35 mg/kg total substance daily for up to 30 days (1–2 weeks beyond confirmed clinical resolution).
  - repeat culture and antibiotic sensitivity may alter antibiotic requirements where there is recurrence or lack of response to empirical treatment.
- topical antiseptics and silver dressings can be used for small areas.
- vaccination is theoretically possible but reserved for where there are several cases in a large stable and where therapeutic progress has not been achieved.
- stable hygiene should be improved.

Prognosis

- re-establishment of the normal cutaneous microbiome and restoring appropriate hygiene measures often leads to dramatic improvement and an improved prognosis.

- cases can be challenging due to clinical pain and difficulty of effective antibiotic delivery.
- good to guarded depending on extent of the infection and sensitivity of organisms.
- underlying conditions and particularly immunocompromised or debilitated horses have a poorer prognosis.

Streptococcal skin infection

Definition/overview

- variety of *Streptococcus* spp. are involved in folliculitis, furunculosis and ulcerative lymphangitis.

Aetiology/pathophysiology

- usually associated with prior skin injury.

Clinical presentation

- small, mildly painful follicular infections occur around the mouth (cheek acne), vulva and wounds.
- may become generalised and/or systemic if *Streptococcus equi* is involved (see page 77, Book 3):
  - lymphadenopathy and lymph node abscess formation is a common feature with painful localised swellings with plasma exudation (Fig. 4.32).
  - secondary immune-mediated (necrotising) vasculitis purpura haemorrhagica can develop 2–8 weeks after infection and cause severe skin destruction, especially of the distal limb regions.

FIG. 4.32 Streptococcal infection involving the lymphatics. There is hair loss over a fulminating abscess, with rupture and discharge of yellow pus, in a typical position under the jowl.

Differential diagnosis

- Staphylococcal folliculitis.
- other bacterial infections.
- dermatophytosis.

Diagnosis

- impression smears or swabs for culture:
  - purulent exudate or moist surface of a crusted lesion.
  - reveals short chains of Gram-positive cocci.
- Reverse transcription (RT)-PCR of skin swabs from intact pustules or moist surfaces of crusts:
  - rapid detection/identification of *Streptococcus equi* subsp. *equi* and other species.

Management

- infected regions can be sanitised with chlorhexidine or povidone iodine washes.
- abscesses drained whenever possible.
- most *Streptococcus* spp. are susceptible or partially susceptible to penicillin:
  - first-line antibiotic for all such infections.
  - 15–20 mg/kg i/m injection daily or twice daily for 5 days is often effective.
  - potentiated sulphonamides (30 mg/kg p/o q24 h) can take up to 4 weeks to resolve these infections.

Prognosis

- most cases recover, but where there is systemic involvement, the outlook is guarded.

Bacterial folliculitis and furunculosis (acne)

Definition/overview

- Folliculitis is inflammation of the hair follicle within the follicular lumen.
- degeneration of the hair follicle leads to infection of surrounding dermis and subcutis (furunculosis) (Fig. 4.33).
- usually hygiene related, with potential for rapid stable spread.
- multiple infection sites may coalesce to form a defined or complex abscess.
- specific conditions relating to bacterial infection of skin occur with:
  - *Staphylococcus aureus* results in often severe pain at the affected site and especially noticed in 'saddle rash' (Fig. 4.34).
  - *Streptococcus* spp. cause of 'cheek abscess' and multifocal pyoderma.
  - *Corynebacterium pseudotuberculosis* causes pigeon breast (Wyoming strangles) (Fig. 4.35).

Aetiology/pathophysiology

- unhygienic skin conditions, complicated by areas of friction from harnesses, rugs or saddle cloths causing injury and allowing infection of hair follicles.
- coagulase-positive *Staphylococcus* spp. are usually involved:
  - cause severe local pain (cardinal sign) and inflammatory responses.

FIG. 4.33 Staphylococcal furunculosis. Note the heavy mat of exudate with hair slough about to occur.

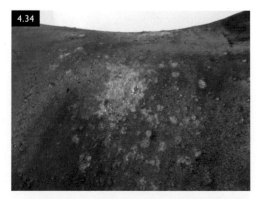

FIG. 4.34 This Thoroughbred racehorse was one of several similarly affected in a racing yard. The horses were in extreme pain. *S. aureus* was isolated from the pustules, and spread was attributed to poor hygiene discipline.

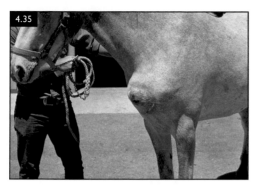

FIG. 4.35 *Corynebacterium pseudotuberculosis* abscess ('Wyoming strangles' or pigeon breast), which can be spread by biting flies. (Photo courtesy AA Stannard.)

- *Streptococcal* spp. infection causes less pain.
- *Dermatophilus* spp. infection causes mild discomfort.

Clinical presentation

- rapidly developing, small, **painful** papules:
  - followed by local oedema and exudation.
  - skin-damage sites:
    - back/saddle region.
    - sides of the chest and girth.
    - pastern and cannon regions.
- increasingly acute painful response and horses may kick/bite during examination.

Diagnosis

- swabs of exudate, pus or deep biopsy for culture/sensitivity testing.
- skin biopsy can confirm the inflammatory response and presence of Gram-positive bacteria in small microabscesses or follicular aggregations.

Management

- control of large areas of infection is challenging.
- pastern dermatitis is also difficult to treat effectively.
- antibacterial washing is essential.
- parenteral antibiotics especially penicillin:
  - can be combined with in-feed potassium iodide.
  - total dose of 3–8 grams daily (weight dependent) for up to 21–28 days.

- surgical drainage of larger lesions may be necessary.
  - followed by parenteral antibiotics:
    - penicillin (24,000 IU/kg i/m q24 h for 3–7 days).
    - trimethoprim/sulphadiazine (30 mg/kg p/o q24 h until 7 days past clinical resolution).
- all in-contact equipment must be washed in hot detergent and sterilised.
- only individualised equipment used.
- hand-washing, sanitisation and wearing of gloves/face masks etc. are essential.
- potentially transmitted between horses and between people and horses and vice versa.

Prognosis

- guarded, with conditions often very slow to resolve, leaving scars and areas of hair loss.
- some horses remain 'sensitive' to skin handling for prolonged periods and liable to reinfection.

Bacterial granuloma (botryomycosis)

Definition/overview

- Botryomycosis (bacterial pseudomycosis) is an uncommon chronic pyogranulomatous condition associated with bacterial infection of skin injuries or some internal structures such as the spermatic cord after castration.
- *Staphylococcus aureus* is the usual cause, but a wide range of Gram-negative organisms have been isolated.

Aetiology/pathophysiology

- infection of lacerations, punctures or a post-surgical complication of skin or other tissues typically by *Staphylococcus* spp. and occasionally *Actinomyces* spp. or *Actinobacteria* spp.
- slowly progresses to multiple miliary interlinking abscesses or microabscesses often discharging through multiple sinus tracts.

Clinical presentation

- retarded healing accompanied by induration of the wound edges.

- chronic purulent discharge from one or more sinus tracts often with rosette formation of granulation tissue around tracts (Fig. 4.36).
- may affect large areas (10–20 cm in diameter) or may be solitary and nodular:
  ○ rupture of the nodules results in discharge of mucoid pus, with whitish yellow granules.
  ○ granules typically contain the infective organism (useful diagnostic material).
- lesions often on the shoulder, neck, withers, ventral abdomen, udder or limbs.
- chronically discharging sinuses at castration sites may progress to Scirrhous cord.
- occasional bacteraemia with disseminated abscessations in internal organs such as lung, liver, spleen and kidney.

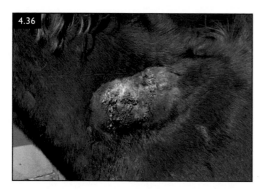

FIG. 4.36 Bacterial granuloma (botryomycosis lesion) showing many discharging sinuses in the granulation tissue.

Diagnosis

- history of the type of injury and slow progression.
- histopathology is characteristic (representative tissue must be collected).

Management

- surgical excision with removal of all infected tissue may be possible:
  ○ brings rapid reduction in bacterial load and positive healing.
- antibiotics depending on culture and sensitivity patterns:
  ○ penicillin (15–25 mg/kg i/m q24 h for 3–7 days).

- oral antibiotics – trimethoprim/sulphadiazine (30 mg/kg p/o q24 h until 7 days past clinical resolution).
- oral potassium iodide – total dose of 3–8 grams daily dependent on size is often helpful.

Prognosis

- guarded.
- reinfection can occur following surgery, leading to repeated procedures.

Dermatophilosis

Definition/overview

- contagious skin infection caused by *Dermatophilus congolensis*:
  ○ Gram-positive, non-acid fast, branching filamentous aerobic, or facultative anaerobic, actinomycete bacterium.
  ○ produces highly resistant zoospores easily transmitted directly or by fomite.
- also affects cattle (often severely) and other farm species.
- characterised by exudation, mild pain and matted hair.
- highest prevalence in warm moist climates.
- affects the dorsum (rain scald) or pasterns (mud fever) in particular.

Aetiology/pathophysiology

- chronically affected (carrier) animals are the primary source of infection.
- zoospores are infective stage of organism:
  ○ require warmth and moisture to become active.
  ○ cause disease through damaged skin:
    ♦ normal healthy skin is relatively resistant to infection.
    ♦ prolonged wetting or dampness of the skin is usually important.
  ○ mechanical transmission of disease via biting/non-biting flies, and possibly fomites.

Clinical presentation

- several different forms of the disorder are recognised in different circumstances.
- **localised or generalised crusting with alopecia** (Fig. 4.37):

FIG. 4.37 Mid-winter (long coat form) of derma-tophilosis after a prolonged rainy period of 10 days. The alopecia is distributed over the water run-off pattern.

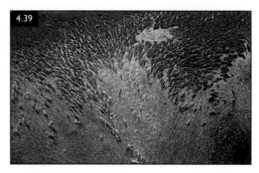

FIG. 4.39 Generalised crusting effect caused by matting of small groups of hair infected with *D. congolensis*.

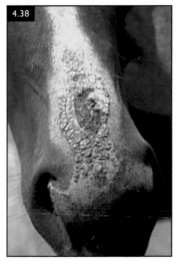

FIG. 4.38 This white-skinned nose shows a combination of signs due to sunburn and infection with *D. congolensis*.

- some cases and sites show severe erythema that may be photo-exacerbated (Fig. 4.38).
- skin atrophy and chronic scarring occur with prolonged infections:
  - more liable to repeated skin infections and may progressively worsen over years.

Winter hair coat form

- thick, creamy, white to yellow or greenish pus accumulates between skin and overlying crust:
  - removal shows slightly raised moist skin with a circular shape.
  - undersurface of plucked crust is often concave and has hair roots protruding through the crust revealing a 'paintbrush' appearance.

Summer hair coat form

- shorter hair means the lesions tend to have smaller crusts (shot-like lumps 1–2 mm in diameter) incorporating 8–10 hairs.
- widespread infection (usually termed 'rain scald') gives a generalised crusting effect (Fig. 4.39).
- excessive and sustained moisture can result in an eczematous state:
  - foals kept in wet unhygienic stables.
  - back of the pastern (greasy heels) in older horses (Fig. 4.40).
- separate clinical entity on the dorsal aspect of the hind cannon bone of racehorses running on cinder tracks or some all-weather surfaces:

- early lesions and crusting often more easily felt than seen in the hair coat.
- modest pain with mild resentment to palpation without pruritus are characteristic.
- **severe cases may show systemic signs:**
  - lethargy, depression, poor appetite, weight loss, fever and possibly enlarged lymph nodes.
- hair is easily pulled out and has typical paintbrush appearance:
  - leaves a silvery, mildly exudative often circular skin surface.
- other lesions harbour a pool of yellowish pus under a haired crust.
- head and limb lesions occur more prominently on white-skinned areas.

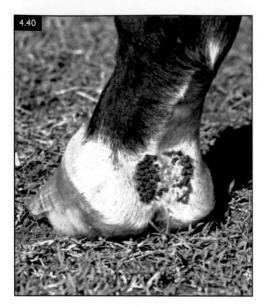

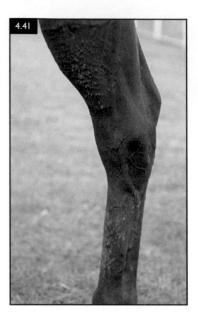

FIG. 4.40 *D. congolensis* was isolated from this horse as the cause of 'greasy heel' associated with alopecia, crusts and cracking of the pastern skin. The condition frequently affects white limbs only.

FIG. 4.41 *D. congolensis* infection in this horse caused small shot-like lesions on the front of the hind cannon caused by working on a cinders-covered training track.

  ○ lesions typical of the summer type, with closely placed, small, matted hair patches down the front of both hind cannon bones (Fig. 4.41).

Diagnosis

- clinical appearance is highly suggestive of the condition:
  - ○ non-specific signs so can be mistaken for other skin conditions of infectious and non-infectious aetiology.
- direct smears from exudate of a fresh lesion stained with usual stains reveals characteristic 'railway line' appearance of organism:
  - ○ microaerophilic bacterial culture is required.
  - ○ samples should be submitted with this in mind.
- fresh lesions cannot be identified:
  - ○ part of a crust macerated with a few drops of sterile water on a glass slide.
  - ○ stained and examined microscopically.
- histopathology of skin biopsies is non-specific but useful.
- RT-PCR is highly sensitive and available commercially:

  ○ positive result should always be correlated with appropriate clinical signs.

Management

- generalised infection in large groups of horses not usually treated because of logistical problems:
  - ○ disease is usually self-limiting.
  - ○ resolution over 3–4 weeks providing wetting of the skin is prevented.
- severe cases – individual treatment is necessary:
  - ○ affected areas of skin gently washed with chlorhexidine and warm water.
  - ○ all infected debris removed from the skin and infected area then kept dry.
  - ○ on white skin areas may crack or fissure:
    - ◆ emollient cream of antibiotics, steroids and zinc sulphate solution.
- concurrent photodermatitis:
  - ○ stabling or protection sun block (SPF30+) of affected area.
- protective bandaging used with caution on lower-limb lesions:
  - ○ may increase severity of the disease by keeping areas moist.

- systemic antibiotic treatment helpful in some cases but seldom necessary:
  - procaine penicillin (22 mg/kg i/m q24 h for 5 days).
  - potentiated sulphonamides (30 mg/kg p/o q24 h for 14–28 days).
- large, generalised skin infections can be treated with:
  - 4% povidone-iodine.
  - chlorhexidine 1–4% wash left *in situ* for 1–15 minutes and then rinsed off.
- remove cases from contact with wet grass or stables:
  - bedding must be clean, non-irritating and dry.
  - skin must be kept dry.

Prognosis

- generally excellent but recurrences are common.
- breakdowns in management or hygiene can result in persistent and severe skin disease especially on white-skinned distal limbs.

Glanders (farcy) (see page 94, Book 3)

Definition/overview

- limited at present to Eastern Europe, Asia, the Middle East and North Africa.
- **notifiable disease.**
- highly contagious, acute or chronic, disease of horses caused by *Burkholderia* (*Pseudomonas*) *mallei* infection (Fig. 4.42).
- both respiratory tract and skin forms are usually fatal.
- **zoonotic** and can be fatal to humans.
- infection usually transmitted by close contact between infected horses, with fomites, and less frequently ingestion of contaminated feed and water.
- rapid ulceration of the skin, leading to discharge of a honey-like secretion

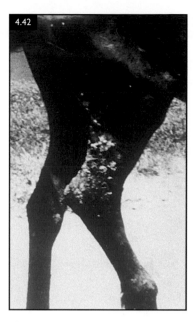

FIG. 4.42 Glanders (farcy). Extreme care should be exercised with all conditions resembling this case. (Photo courtesy AA Stannard; reprinted from Knottenbelt DC and Pascoe RR (1994) *Colour Atlas of Diseases and Disorders of the Horse*, Mosby, with permission.)

associated with corded multinodular lymphatics, lymphangitis and regional lymphadenopathy.
- confirmed by intradermopalpebral (mallein) test, culture and/or histopathology:
  - other tests include:
    - PCR.
    - various serological tests:
      - complement fixation test (CFT), immunofluorescent antibody (IFA) test, enzyme-linked immunosorbent assay (ELISA).
- major differential epizootic lymphangitis (histoplasmosis).
- euthanasia is always recommended due to contagious nature and zoonotic potential.
- many countries have elimination policies.

COMMON FUNGAL SKIN INFECTIONS

Dermatophytosis (ringworm) caused by *Trichophyton* spp.

Definition/overview

- common highly contagious disease affecting horses of all ages.
- fungal infection of the hair follicles caused by *Trichophyton* spp.:
  - *T. equinum* var. *equinum* mostly in northern hemisphere.
  - *T. equinum* var. *autotrophicum* predominately in the southern hemisphere.
  - *T. verrucosum* and *T. mentagrophytes* also occur.
  - organisms are generally species-specific but cross-infections do occur.
- spread by fomite or direct or indirect contact with the highly resistant spores.
- widespread, rapidly spreading infections are common in stables with poor hygiene.

Aetiology/pathophysiology

- organisms are found on horse's hair or fomites, and rarely in soil.
- infection requires minor skin trauma contaminated with the organism.
- clinical disease 9–15 days after infection.

Clinical presentation

- loss of hair and scaling of skin.
- initially hairs become erect in a circular area 5–20 mm in diameter (Fig. 4.43).

- actively infected sites usually variably pruritic and occasionally painful.
- initially single lesions and then infected area covered with multiple, often coalescing lesions (Fig. 4.44).
- hair plucks easily 10–15 days post-infection leaving a nearly hairless, silvery surface:
  - usually, no pain associated with plucking.
  - marked contrast to infections due to *Microsporum* spp.
- may result in a diffuse crusting dermatitis over area (cf. pemphigus foliaceous):
  - common in girth region (Girth Itch).

Diagnosis

- hairs plucked from the margin of a fresh lesion previously wiped with alcohol (to reduce bacterial and mould contaminants):
  - submitted for culture and direct microscopic examination.
  - direct smears often reveal hyphae and large endothrix spores.
- *Trichophyton*-infected hairs do not fluoresce under black UV light.
- culture of infected hair on Sabouraud's agar or rapid sporulating medium (RSM) and dermatophyte test medium (DTM) at 25°C (77°F):
  - diagnosis but very slow – often weeks.
- skin biopsies/hair samples submitted for PCR analysis gives fast and reliable speciation.

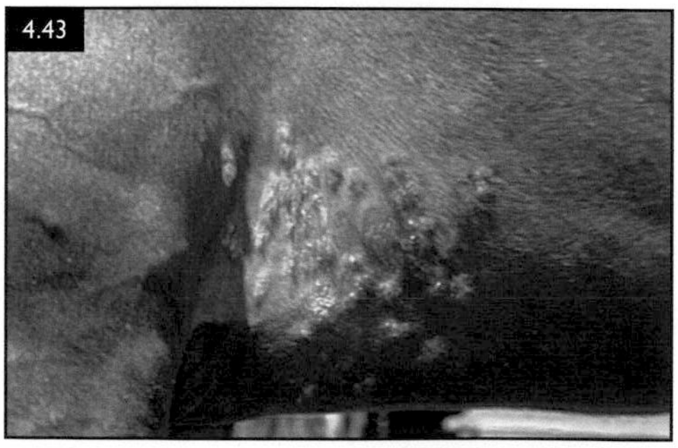

FIG. 4.43 Early infection with *T. equinum* var. *equinum* due to the use of a contaminated girth 14 days earlier is seen as small areas of alopecia and small papules with erect hairs.

FIG. 4.44 An older dermatophyte infection with a large area of alopecia. The early lesions have coalesced, with a secondary infected area of smaller peripheral lesions.

- cutaneous biopsies usually reveal endothrix and mural folliculitis in 80% of cases.

Differential diagnosis

- Dermatophilosis.
- *Microsporum gypseum* infection.
- Sweet itch   • insect hypersensitivity.
- mite infestation.
- Equine sarcoidosis   • sarcoid.
- Multisystemic eosinophilic epitheliotraopic disease.
- alopecia areata.
- Anhidrosis
- actinic dermatitis   • wound scalding.
- *Malassezia* spp. infection.

Management

- most horses develop resistance to dermatophytes and infection is usually self-limiting over 4–8 weeks.
- control is achieved by preventing spread between horses:
  ○ all crusts and infected hairs should be carefully removed and burnt.
  ○ all tack and infected buildings fumigated or sprayed with a halogenated peroxygen or a quaternary ammonia disinfectant to prevent spread.
  ○ infected horses handled last and strict post-contact hygiene instituted.
- sprays, washes or ointments containing miconazole or enilconazole, natamycin, or terbinafine are usually used.
- Fluconazole (5 mg/kg q24 h) is an effective systemic treatment.
- 'Ringworm vaccination' developed for cattle using *T. verrucosum* widely used and developed into a commercial equine vaccine:
  ○ useful as part of a holistic control system in widespread outbreaks.
  ○ can cause significant swellings and even abscessation at injection site.
  ○ not available in all countries.

Prognosis

- good for individuals.
- long-term/permanent hygiene and quarantine measures should be used to restrict spread.

Ringworm caused by *Microsporum* spp.

Definition/overview

- relatively common fungal infection of the hair follicles caused by *Microsporum* spp.
- common species causing equine dermatophytosis are:
  ○ *M. gypseum* a soil saprophyte.
  ○ *M. canis* typically affects dogs and cats.
  ○ occasionally *M. equinum*.

Aetiology/pathophysiology

- usually spread by contact with contaminated area (e.g. transport, tack, soil, blankets):
  ○ also spread by biting insects and localised skin abrasion.
- microspores invade the outer hair shaft, causing ectothrix infection.
- infection does not destroy all the hair in the infected area:
  ○ leaves a moth-eaten look to the hair coat.

Clinical presentation

- small, usually well-defined, hairless areas (alopecia) commonly on the face and limbs:
  ○ lesions may follow a distributed pattern of insect bites (Fig. 4.45).

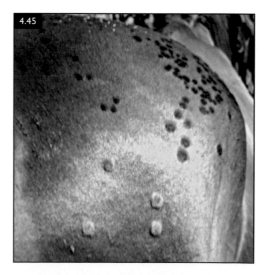

FIG. 4.45 *M. gypseum* spread by *Stomoxys calcitrans* flies. The distribution of infected sites corresponds to areas where *S. calcitrans* has fed on the horse.

- not all hairs are affected/shed:
  - hair plucking usually painful (cf. *Trichophyton* spp.).
- secondary bacterial infection relatively common:
  - purulent discharges may be present.
- uncomplicated active lesions are typically not pruritic:
  - contrast to *Trichophyton* spp.

Differential diagnosis

- Trichophytosis
- *Stomoxys* bites.
- sweet itch.
- mites
- nymph ticks
- lice.
- sarcoid.
- *Malassezia* spp. infection.
- some bacterial infections of the skin.

Diagnosis

- hairs plucked from a fresh lesion examined for ectothrix:
  - spores/fungal hyphae usually seen by direct microscopy or histopathology.
- Wood's light causes fluorescence of only *M. equinum* and some *M. canis*.
- culture of hair on Sabouraud's agar or RSM and/or DTM at 25°C (77°F).
- hairs from periphery of lesions submitted for PCR amplification:
  - dermatophyte genus/species identification available at some laboratories.
- skin punch biopsies from the periphery of affected areas:
  - may reveal ectothrix and mural folliculitis in 80% of infected submissions.

Management

- treatment is per *T. equinum* infection (see page 94):
  - response is usually slow and variable, but lesions heal with time.
  - most cases self-cure with a reasonable seroconversion over 8–12 weeks.
  - immunity lasts around 1 year.
- all affected horses should be quarantined.
- *M. canis* cultured or identified on PCR:
  - identify and treat asymptomatic carrier cats, dogs or rodents.
- Fluconazole (5 mg/kg q24 h) effective systemic treatment for ringworm in horses.

- Griseofulvin is probably not useful.
- control measures as described for Trichophytosis.

Prognosis

- good.
- very difficult to eliminate the condition from a premises.
- recovery is slower than *Trichophyton* spp.

Malassezia spp. dermatitis

Definition/overview

- almost all cases are opportunistic, secondary infections.
- yeast may be isolated from approximately 50% of normal horses:
  - interpret the significance of culture with care.
- intensely pruritic, erythematous skin condition.

Aetiology/pathophysiology

- often related co-morbidity with systemic immunocompromising disorders or prolonged corticosteroid use.
- signs may be due to hypersensitivity to environmental allergens or the organism itself.

Clinical presentation

- varies from erythematous, dry and scaly patches to greasy areas with brown-black exudate.
- often found in areas that trap heat and moisture:
  - intermammary region and preputial fossa.
- intensely pruritic, with self-trauma to local areas including:
  - tail.
  - ventral midline including the intermammary region in mares.
  - pruritus often poorly responsive to glucocorticoid therapy.

Diagnosis

- skin swabs and/or direct impression smears:
  - may reveal clusters of yeasts adherent to keratinocytes.

- culture attempted on Sabouraud's chloramphenicol agar if confirmation essential.

Management

- underlying primary conditions should be resolved:
  - normal hygiene procedures instituted.
  - often leads to resolution.
- keratolytic degreasing shampoo followed by anti-yeast agents:
  - chlorhexidine with either miconazole, climbazole, clotrimazole, enilconazole or posaconazole.
  - shampoo, spray, mousse or ointment applied directly to the skin once daily until the condition resolves.
- anti-inflammatory therapy, i.e. glucocorticoids, used with caution to avoid further immunocompromise.

Prognosis

- favourable for the condition but remission dependent on control of underlying conditions.

Histoplasmosis (epizootic lymphangitis [EZL])

Definition/overview

- rare outside its geographical restriction, but notifiable disease.
- chronic, contagious disease characterised by suppurative lymphangitis.
- caused by *Histoplasma capsulatum* var. *farciminosum.*
- eliminated in Europe but endemic in Saharan Africa, Middle East and Far East.
- affects horses mostly, but donkeys and mules also affected.
- affects the skin on the limbs:
  - also, neck, lips and areas of harness abrasions.
  - pulmonary and ophthalmic forms also recognised.
- asymptomatic carriers play a part in the epidemiology.
- potentially zoonotic

Aetiology/pathophysiology

- caused by fly-borne *Histoplasma capsulatum* var. *farciminsum.*

- insidious onset with variable incubation (weeks to many months).

Clinical presentation

- severe, chronic progressive debilitating disease primarily affecting the skin:
  - chains of suppurative, ulcerating, pyogranulomatous cutaneous nodules.
  - nodules may ulcerate over time producing a thick purulent material laden with fungal organisms.
- early cases are subtle: retain appetite and normal weight.
- progressive involvement causes debility, lameness and other signs of nasal, pulmonary and ocular/conjunctival involvement.

Differential diagnosis

- Glanders (*Burkholderia mellei*).
- *Sporothrix schenkii* (sporotrichosis).
- fungal granuloma especially black grained mycetoma (*Curvularia geniculata*).
- Ulcerative lymphangitis (*Corynebacterium* spp.).
- Strangles (*Streptococcus equi*).

Diagnosis

- clinical signs and geographical location.
- direct smears from discharges are pathognomonic:
  - culture (very slow) and PCR methods are available.
- skin 'Histofarcin' test can help detect early cases.

Management

- early cases managed by incision of nodules, expression of pus and repeated irrigation of ulcerated nodules with strong iodine solution.
- all material from treatment disposed by burning.
- intravenous infusions of sodium iodide or prolonged oral potassium iodide (20 mg/kg) can help some cases.
- amphotericin B can be used but is usually impractical/too costly.
- oral antifungal agents are largely ineffectual.
- occasional cases appear to overcome the infection.

- advanced cases euthanised and disposed by burial or burning.

Prognosis

- very guarded but determined efforts can bring satisfactory resolutions.
- no vaccine is available.

Alternariasis

Definition/overview

- relatively common, usually minor granulomatous skin disease.
- caused by intradermal growth of *Alternaria* spp. fungal organisms.

Aetiology/pathophysiology

- *Alternaria* spp. organisms transferred by biting flies:
  - *Simulium* spp./sand flies largely responsible.
- slowly expanding, non-painful, firm discrete granuloma lesions usually restricted to the skin.
- free-living saprophytic fungus growing in rotting vegetation, dung heaps and composted vegetation.
- other opportunistic fungal organisms may be involved in similar pathology.

FIG. 4.46 6-year-old Irish Sport Horse developed a few similar dark, pain-free nodules in the skin of the ears and across his face. Biopsy confirmed a fungal granuloma assumed to be *Alternaria* spp.

Clinical presentation

- most lesions occur on the ears (Fig. 4.46):
  - either inside or outside skin surfaces.
  - majority on ear extremities.
- other sites affected include the back and neck.
- one/few, solid, discrete, well-defined nodules:
  - usually, 3–5 mm diameter but can be much larger.
  - non-painful and asymptomatic.
- some cases become more severe and other body sites affected as well:
  - variable areas of coalescent hairless nodules.
  - may be painful on palpation.

Differential diagnosis

- sarcoid
- skin cysts.
- collagen granuloma.

Diagnosis

- clinical signs.
- histology is typical.
- specific identification of organism is challenging.

Management

- surgical removal is curative.
- can be left untreated but may have a cosmetic consequence.
- large lesions and those in challenging sites (dorsal back region):
  - oral systemic potassium iodide at 10–40 mg/kg body weight repeated daily in feed for up to 30 days.
  - **note: do not administer potassium iodide to a pregnant mare.**

Prognosis

- excellent since limited clinical effects.
- large areas are more problematic.

Pythiosis (phycomycosis/ Bursatti/Florida leeches/ swamp cancer)

Definition/overview

- subtropical and tropical chronic, subcutaneous, ulcerative and granulomatous skin disease caused by *Pythium* spp. especially *insidiosum*.

FIG. 4.47 *P. insidiosum* lesion on a horse's belly. Note the thick stringy exudate. The horse exhibits severe irritation, with biting and rubbing of the affected area.

FIG. 4.48 Pythiosis in a chronic limb lesion. Note the characteristic numerous discharging, granulomatous lesions, with a stringy serosanguineous exudate.

- usually occurs in summer and autumn.
- all breeds, ages and sexes:
  - dark coat colour and darkly pigmented skin are predisposed (insect preference).

Aetiology/pathophysiology

- usually requires damaged or macerated skin to assist entry of the organism:
  - long periods standing in stagnant water containing rotting organic material at high ambient temperatures (30–40°C).
  - blood-feeding insects may carry the organism and cause micro-wounds.

Clinical presentation

- progressive development of sticky, serosanguineous, stringy discharges with hair matting (Figs. 4.47, 4.48):
  - hang from body wall and/or limbs in thick mucopurulent strands.
- often severely pruritic with biting and kicking at affected areas:
  - repeated trauma and exudate.
- large (1–2 cm) aggregations of irregularly shaped, yellow-tan to grey, gritty structures ('kunkers', 'leeches') found buried in the fibrous tracts.
- lymphadenopathy in most cases.
- involvement of joints/tendons with sinus formation represents a serious complication.

Diagnosis

- geographically restricted and suspected when access to waterlogged pasture/lagoons.
- serodiagnosis ELISA antibody test.
- early lesions biopsied:
  - fresh tissue or exudate samples cultured on selective media.
  - histopathological findings include eosinophilic granulomatous inflammatory reaction with intralesional hyphae characteristic of *Pythium insidiosum*.
  - immunohistochemical assays developed for use with formalin-fixed tissues.
- real-time PCR for detection of organism provides a rapid and reliable diagnosis.

Management

- spontaneous remission does not occur, and early treatment is always required.
- repeated surgical excision under GA:
  - most successful treatment, particularly in chronic cases.
- sodium iodide (7 mg/kg i/v as a 3.5% solution in normal saline repeated at 7 days):
  - useful adjunctive effects and may reduce the size of lesions prior to surgery.
- other systemically administered antifungal drugs can help but are not usually curative due to organism having some resistance to antifungal agents.

- phenolised vaccine has shown some success in early cases.
- distal limb lesions respond well to surgical excision of exuberant granulation tissue:
  - followed by intravenous regional limb perfusion of a solution containing:
    - ◆ 50 mg (in 10 ml) of amphotericin B.
    - ◆ 6 ml of medical-grade DMSO.
    - ◆ 44 ml of lactated Ringer's solution.

Prognosis

- guarded in all cases but dependent on size and site of the lesion, duration of infection, immunocompetence and available treatment.
- body-wall lesions better prognosis than distal limb lesions.

PARASITIC DISEASES

Ectoparasites

Pediculosis (lice)

Definition/overview

- *Werneckiella (Damalinia/Bovicola) equi* chewing lice:
  - 1–2 mm in length, reddish brown or beige.
  - usually found on head, neck and dorsolateral trunk (Fig. 4.49).
  - feed on skin scale and exudates.
- *Haematopinus asini* sucking lice (Fig. 4.50):
  - 2–3 mm and are found on the body, limbs, mane and base of the tail.
  - feed on blood and tissue fluids.
- both often associated with poor management or body condition:
  - immunocompromised horses may have heavy infestations under good management conditions.
- both species cause variably intense irritation, dermatitis and general unthriftiness:
  - blood-feeding lice more obvious systemic effect.
- **common causes of pruritus occurring during colder months.**

Aetiology/pathophysiology

- highly contagious causing widespread outbreaks in stables.
- cannot survive off the host for more than 3 weeks.
- eggs and parasites are easily spread through direct contact or by grooming equipment, rugs and harness etc.
- heavy infestations are most common in winter.
- *Haematopinus* is more common in donkeys and more resistant to higher temperatures.

Clinical presentation

- light infestations are often asymptomatic.
- clinical signs mainly relate to self-trauma:
  - restlessness, rubbing and biting with broad areas of damaged hair.
  - self-inflicted physical injury to the skin with serum exudation.

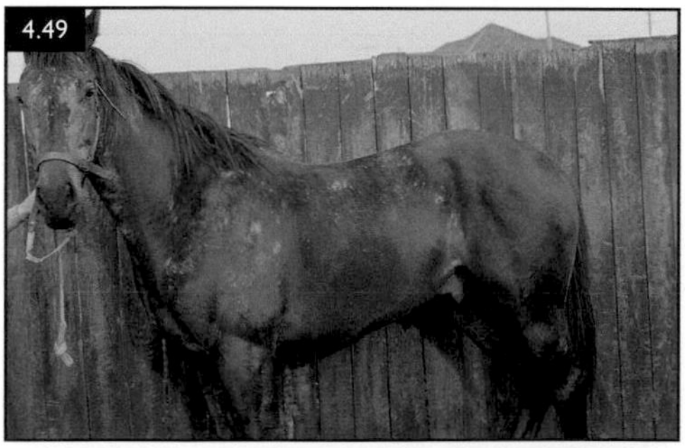

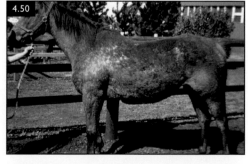

FIG. 4.49 Pediculosis. *Werneckiella (Damalinia) equi* on a 4-year-old racehorse with patchy alopecia, loss of head hair from rubbing, some scurf and severe pruritus.

FIG. 4.50 Pediculosis. *Haematopinus asini* all over the body, with alopecia and a mildly seborrhoic greasy skin in a long-haired winter coat.

- *H. asini* does induce primary skin damage (pinpoint haemorrhage).
- loss of body condition in unhygienic or poorly managed housed horses.
- heavy infestation with *Haematopinus asini* may cause anaemia.

Diagnosis

- parasites are visible to the naked eye:
  - heavy infestations, parasites are found over the entire horse.
  - warming the skin with rugs can bring the parasites to the surface.
- skin brushings/combings readily identify the parasites.
- eggs ('nits') usually visible with magnification as pale-yellow specks attached firmly to hair shafts.
- *W. equi* may be present in small numbers and can then be difficult to find.
- *H. asini* usually present in larger numbers.
- lice and eggs identified by microscopic examination.

Management

- shedding winter hair coat and sunshine exposure can reduce the burden rapidly.
- contact between horses, and horse equipment, should be restricted.
- parasites and eggs on equipment easily killed by heat (steam cleaning).
- *H. asini* treated efficiently with oral ivermectin (0.2 mg/kg twice at 14-day intervals).
- *W. equi* do not respond to oral avermectins.
- Fipronil, permethrin and cypermethrin powders, washes and sprays can be used for both species and applied weekly.
- 1% selenium sulphide shampoo or lime-sulphur washes also effective.
- **manufacturer's instructions, dosage and concentration must be followed for all sprays, washes and powders.**

Prognosis

- excellent outlook for all cases of pediculosis.
- usually self-limiting with hot weather and suitable hygiene measures.
- immunocompromised horses require a different approach which may affect the prognosis.

Chorioptes equi infestation (chorioptic mange/leg itch)

Definition/overview

- *Chorioptes equi* is a surface-feeding parasitic mite that causes pruritus and consequent dermatitis.
- most often affects the lower legs of feathered horses.
- non-feathered horses can also be affected, and wider infestation is possible.

Aetiology/pathophysiology

- *C. equi* feed on epidermal debris and they remain on the surface of the skin.
- movement causes intense itchiness/pruritus in affected horses.
- life cycle lasts around 3 weeks with all stages occurring on the host.
- mites and their eggs can survive up to several weeks off the host:
  - opportunities for transmission, reinfestation and spread within a stable yard.
- transmission by direct and indirect contact:
  - more common in stabled conditions and particularly with straw bedding.
  - parasites can be transferred by grooming brushes etc.
- lower limbs are mainly affected but can be more generalised.
- carrier cases harbour the infestation over the summer months and often show little or no pruritus.

Clinical presentation

- pruritus with irritation and restlessness especially at night and in the stable:
  - degree of pruritus does not reflect the level of infestation.
  - stamping/biting at the limbs causes superficial dermatitis and erosion/ulceration.
  - rubbing on rails, fences and posts often reported when parasites more generalised.
- skin damage may result in serum exudation and matting of limb hair in feathered horses with frequent secondary bacterial infections.
- hair damage ultimately leads to baldness and progressive crusting (Fig. 4.51).
- chronic long-standing cases can cause:

FIG. 4.51 Chorioptic mange in a draught horse, presenting with severe irritation, stamping, rubbing and biting at the lower limbs. The hair on the back of the limbs has been clipped to reveal raised crusts and areas of dried exudate.

- ○ very aggressive epidermal hypertrophy and fibrosis from repeated self-trauma.
- ○ skin becomes folded and intertrigo dermatitis can occur:
  - ♦ self-perpetuating chronic skin condition with exudation and fibrous proliferation.

Diagnosis

- clinical signs and history are typical for limb forms of the disease but whole-body cases are more challenging.
- parasites are usually easily collected using skin brushings with a stiff bristle brush:
  - ○ careful microscopic examination of dander required immediately after collection.
  - ○ magnifying option or app on smartphone camera very useful stable side.
  - ○ mites easily recognised, slow-moving and in large numbers.

Management

- topical treatments are labour intensive but currently the most effective tool:
  - ○ feathers clipped to allow adequate penetration of topical medications:
    - ♦ may be resisted, especially for show horses (up to 2 years for return).
  - ○ topically applied macrocyclic lactones address surface-feeding of the mites.
  - ○ off-label use of eprinomectin pour-on solution (500 µg/kg body weight) sprayed on affected areas once weekly

for four applications appears an effective and safe alternative.
  - ○ other options include:
    - ♦ topical organophosphates – malathion (0.5%) and coumaphos (0.06%).
    - ♦ topical permethrin.
    - ♦ selenium sulphide shampoo followed by lime sulphur (37 g/l), sponged on every 5 days for 1 month.
    - ♦ fipronil spray (0.25%).
- systemic therapy may be an option though efficacy is suboptimal:
  - ○ Ivermectin at 0.2 mg/kg p/o weekly for 4 doses.
  - ○ Doromectin repeated intramuscular injection of 0.3 mg/kg body weight at 2-week intervals for up to 6 procedures.
- all in-contact animals and affected horses treated simultaneously.
- environmental decontamination is important, including barn, stalls and bedding, tack and grooming equipment:
  - ○ steam cleaning and sterilisation of all harness, grooming equipment etc.
- carrier animals should be detected out of season and treated effectively before being brought inside.
- early detection of a problem in a stable yard is essential.

Prognosis

- depends heavily on the compliance of all parties involved in the stable.
- all horses should be treated or tested or both.
- recurrence is common.

Poultry red mite infestation

Definition/overview

- *Dermanyssus gallinae* is a poultry mite that emerges at night to feed on the host.
- easily recognised and becomes bright red after feeding.
- horses can be severely attacked when infested poultry or pigeons are in the same environment.
- mites can survive for up to 4–5 months without feeding.
- humans, dogs and cats also affected.

FIG. 4.52 *Dermanyssus gallinae.* A 5-week-old foal is lying down and biting its limbs. Severe pruritus and self-mutilation are common with poultry red mite infestation.

Aetiology/pathophysiology

- blood-sucking mite that lives and lays egg in cracks in the walls and ceilings of poultry houses and pigeon nesting premises.
- life cycle spent mostly away from host animals and birds:
  ○ as little as 7–10 days thereby generating vast populations.
- all equine infestations have at least some poultry or pigeon contact.

Clinical presentation

- severe pruritus, irritability and irritation, leading to stamping and biting of limbs and body (Fig. 4.52).

Diagnosis

- clinical signs and circumstances.
- small red mobile mites may be observed in brushings taken from an affected area:
  ○ sampling best at night.
  ○ populations increase over winter and decrease over summer.
  ○ sampling from affected birds is a useful addition:
    ♦ mites usually found on poultry more easily than on horses.
- careful microscopic examination of shallow skin scrapings to identify mites.

Management

- spray horse with 0.25% malathion.
- removed from proximity to poultry, or poultry removed.
- environmental measures are essential:

  ○ poultry sheds should be power sprayed with 0.5% malathion.
- treatment repeated in 7 days in summer and 10–14 days in winter.

Prognosis

- good if spray is used effectively and repeatedly to eliminate mites.

Psoroptic mange

Definition/overview

- *Psoroptes* spp. mites cause pruritus, patchy alopecia and excoriation primarily restricted to the head and ears.
- aural infestation leads to 'cornflake' appearance on concave surface of the ear canal:
  ○ possible side-to-side head-shaking signs.
- tail-rubbing in young horses.

Aetiology/pathophysiology

- *Psoroptes equi* is a large mite that feeds on tissue fluids.
- 10-day life cycle and can survive off the host for up to 6–12 weeks.
- direct and indirect transmission occur.
- contact with infested young stock or contaminated stables or tack.
- mites most active in areas with thick denser hair such as the ear, mane and tail.

Clinical presentation

- pruritus and irritability around the head, leading to head rubbing, crusting and discharge.
- itchiness of tail, mane and body in stabled yearling horses.
- some horses display 'droopy' ears (Fig. 4.53).

Diagnosis

- examination of ear wax with a hand lens or microscope may reveal small, white, moving parasites.

Management

- sedation may be necessary to clean all accumulated wax and debris from the ears.
- body and tail itching in yearlings respond well to:
  ○ malathion washes (50–125 ml/10 litres).

FIG. 4.53 Psoroptic mange in a stabled horse, showing floppy ears due to the presence of *P. equi* mites in the ears.

FIG. 4.54 16-year-old coloured mare was presented with extreme pruritus of 5 months duration that had not responded to any standard measures. The mare was severely immunocompromised by undiagnosed lymphoma and was euthanised on human grounds.

- ○ oral ivermectin (0.2 mg/kg) or oral moxidectin (0.4 mg/kg) repeated at least twice at 14-day intervals.
- ○ off-label use of topically applied eprinomectin pour-on solution:
  - ◆ 500 µg/kg by spray onto affected areas.
  - ◆ once weekly, four times is more effective and safer.

Prognosis
- excellent with determined efforts and all affected horses are identified and treated.

Sarcoptic mange

Definition/overview
- burrowing mite *Sarcoptes scabiei* var. *equi.*
- lesions mostly on the head, neck and ears but can extend over the entire body.
- **most affected animals are profoundly immunocompromised.**
- rare disease but reportable in many countries.

Aetiology/pathophysiology
- mites tunnel into the skin and feed on tissue fluid and epidermal cells.
- life cycle of 2–3 weeks.
- cannot survive off the horse for more than 72 hours (susceptible to dehydration).
- spread by direct and indirect contact.

Clinical presentation
- intense pruritus with papules and vesicles on the skin (Fig. 4.54):

- ○ often exacerbated by warmth including rugs and warm ambient temperatures.
- rubbing and biting of limbs and ventral abdomen leads to self-inflicted excoriation:
  - ○ skin becomes very crusty and thickened (lichenification).
- secondary bacterial infections are common, especially if immunocompromised.

Diagnosis
- multiple deep skin scrapings and biopsies submitted for microscopic identification of *S. scabiei* var. *equi.*

Management
- oral ivermectin (0.2 mg/kg at 14-day intervals for at least 4–6 treatments).
- oral moxidectin (0.4 mg/kg at 14-day intervals).
- 0.1% diazinon spray every 7–10 days for 3–6 treatments.
- off-label use of topically applied eprinomectin pour-on solution:
  - ○ 500 µg/kg sprayed on affected areas once weekly for four applications.
  - ○ more effective and safer alternative.
- all infected horses isolated, and all potential fomites fumigated.

Prognosis
- largely depends on health status of horse in general:

o debilitated or immunocompromised horses carry a poorer prognosis.

Trombiculidiasis

Definition/overview

- *Trombicula (Neotrombicula) autumnalis:*
  o harvest mites, American chiggers or scrub itch mites.
- mostly affect head and limbs of pastured horses grazing chalky alkaline soils:
  o often geographically restricted.
- stabled horses may be infested over the entire body through contact with infested pasture hay used as feed or bedding.

Aetiology/pathophysiology

- parasites are strongly seasonal, tending to breed in late summer and autumn.
- mites and nymphs are free-living in pastures:
  o only larval stage is parasitic.
  o can remain on the skin for up to 24–36 hours.
- can attack horses and humans if given the opportunity.
- usually feed on horses in the late afternoon resulting in pasture-related regionally restricted pruritus and skin inflammation.

Clinical presentation

- limb stamping and lower limb biting, with nose-rubbing:
  o small orange papules on the limbs, nose and body.
- small hairless areas develop after 2–3 days (Fig. 4.55).

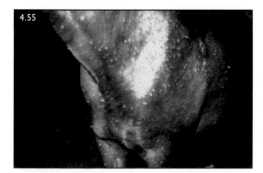

FIG. 4.55 Trombiculid mites bite horses at pasture, causing hairless inflamed areas usually around the head and limbs.

Diagnosis

- confirmed by identification of mites by direct microscopy:
  o mites fall off immediately after feeding, making diagnosis more difficult.
- multiple scrapings can be taken of the most recent lesions.

Management

- Fipronil, permethrin, lime sulphur or 0.25% malathion as a limb and body spray can be used.
- 0.5% malathion can be sprayed onto walls, floors and sand rolls.
- treatment only necessary if severe infestation present or unavoidable environmental conditions mean that isolation for a week or two is not possible.

Prognosis

- excellent and can be prevented by careful management of the grazing environment.
- seasonally related recurrences are common.

Larval nematode dermatitis

Definition/overview

- larval forms of *Pelodera strongyloides* and *Strongyloides westeri* can cause pruritus on the limbs of horses.
- free-living larvae invade the skin of the lower limbs, particularly foals in wet weather.
- occurs under conditions of poor hygiene or contaminated bedding, especially straw.
- marked to moderate pruritus, stamping feet, restlessness and frenzied activity.
- papules, pustules, ulcers, crusts, alopecia, erythema and swelling of the limbs, ventral thorax and abdomen:
  o secondary self-inflicted traumatic dermatitis.
- diagnosis by skin scrapings reveal motile nematode larvae on microscopy.
- skin biopsy shows perifolliculitis, folliculitis and furunculosis with nematode segments present in hair follicles.
- remove all animals from contaminated yards, change bedding, wash the skin with

warm water, and possibly apply topical skin creams (antimicrobial).
- more severe (usually secondary) clinical signs regress over 3–4 weeks.
- therapeutic options for persistent lesions:
  - lime-sulphur washes (warm 2–3% solution) applied twice weekly at least four times.
  - ivermectin at 0.2–0.4 mg/kg (200–400 µg/kg) orally once weekly four times.
- antibiotics may be necessary for concurrent secondary bacterial pyoderma.
- prognosis is excellent.

Onchocercal dermatitis

Definition/overview

- *Onchocerca reticulata [cervicalis]* and occasionally other species including *O. gutturosa* and *O. lienalis*. May cause cutaneous and ophthalmic disease.
- microfilaria can be found in the skin of normal horses.
- hypersensitivity appears to develop in some horses affecting the skin of the ventral midline, chest, withers, face and neck.

FIG. 4.56 *Onchocerca* infection, showing typical head lesion of 4 weeks' duration with scaling, crusts and alopecia. The horse rubbed the affected areas frequently.

Aetiology/pathophysiology

- *Culicoides* spp. intermediate hosts for *O. cervicalis*.
- *Simulium* spp. for *O. gutturosa* and *O. lienalis*:
  - adult parasites inhabit the ligamentum nuchae and the flexor tendons of the limbs causing diffuse non-painful swellings with skin usually remaining intact.
- microfilaria circulating in superficial layers of the dermis cause localised pruritus:
  - may also affect conjunctiva and possibly the uveal tract (see Chapter 2, Book 5).
- dermatitis associated with types 1 and 3 hypersensitivity to microfilarial antigen.
- multiple severe infestation may result in discharging sinus tracts, especially at the withers and neck sites.
- marked seasonality associated with the vector midges.

Clinical presentation

- usually, multifocal painless cutaneous swellings, occasionally with ulceration.
- localised pruritus and patchy alopecia:
  - small papules and thickened, dry, scaly skin (Fig. 4.56).
  - focal pain particularly associated with withers or neck regions.
- severe cases show marked pruritus with secondary excoriation:
  - crust formation ('bull's-eye' lesion on the skin).
- ventral midline form shows severe alopecia and poor regrowth of new hair (Fig. 4.57).
- tail-rubbing is rare but there can be loss of mane or tail hair.
- severe chronic infestation can cause a fistulous condition associated with the nuchal ligament at the withers.

Diagnosis

- easily mistaken for other causes of focal swelling and pruritus.
- skin biopsy usually reveals larval worms in the skin.
- probing of discharging sinus may detect adult parasites:
  - very long, fine cotton-thread-like parasites.

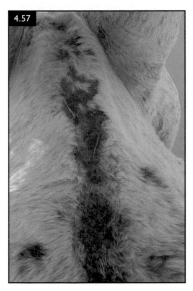

FIG. 4.57 Ventral midline dermatitis due to *Onchocerca* infection. There is loss of hair and scaling, with scattered lesions either side of the midline, and mild to severe pruritus with biting and rubbing of the thickened belly wall.

Management

- Ivermectin (0.2 mg/kg) or moxidectin (0.4 mg/kg) given orally and repeated at 2-month intervals:
  - most cases resolve with a single treatment.
- massive simultaneous death of adult worms or microfilaria can exacerbate the eye and skin inflammation for 3–5 days after treatment:
  - concurrent anti-inflammatory glucocorticoid therapy may help to minimise hypersensitivity reactions to the dying parasites.

Prognosis

- good although some changes may persist in the ligaments and tendons affected.

Oxyuris (equine pinworm/rat tail)

Definition/overview

- Adult *Oxyuris equi* worms inhabit the caecum, large and small colon, and rectum:
  - attach to the gut wall but do not cause any damage or feed.
- infection is stable- or yard-acquired.

Aetiology/pathophysiology

- adult females emerge from the anus and lay rafts of eggs (up to 50,000) on the perineum and perianal skin.
- causes marked and often severe irritation and self-induced local damage in the perineal region.
- tail movement dislodges the eggs which drop into the bedding:
  - very sticky, adhere readily to the lips and then swallowed in huge numbers.
  - prepatent period is variable from weeks to 5–6 months.
- sticky rafts of eggs easily transferred between stables by boots, barrows and forks etc.

Clinical presentation

- perineal pruritus, which can be severe, with widespread local ulceration and secondary infection.
- horses rub repeatedly at the anal or buttock regions:
  - seldom rub over the base of the tail.
- heavily infested horses may lose body condition and become irritable and difficult.

Diagnosis

- clinical signs including rafts of eggs visible as yellow streaks and spots around the anus.
- eggs on perineum skin can be harvested with adhesive tape or scrapings and are pathognomonic.
- faecal samples do not show these eggs, but adult worms may be found (characteristic shape).

Management

- stable/yard hygiene is fundamental.
- all affected horses identified and simultaneously treated.
- regular deworming is theoretically effective but avermectins are not usually helpful.
- Fenbendazole or pyrantel are more effective:
  - high doses ensure adequate therapeutic dose is present in terminal small colon.
- older compounds such as piperazine or thiabendazole are also effective.

Prognosis

- treatment can be problematic, but persistence is usually rewarded.
- essential that hygiene measures are maintained.

Habronemiasis

Definition/overview

- seasonal and geographically limited sporadic ulcerative cutaneous dermatitis:
  - deposition of third-stage larvae of *Habronema muscae*, *H. majus* and *Draschia megastoma* by flies on wounds and other moist areas (prepuce and conjunctiva).
  - Dipteran intermediate fly hosts including *Stomoxys*, *Musca* and *Hematobia* are essential.
- affected horses and donkeys often more severely affected in sequential years.

Aetiology/pathophysiology

- cutaneous signs probably partially result of hypersensitivity response to larval forms.
- geographical location.
- poor hygiene measures controlling fly population:
  - poor manure collection or disposal.
  - moist patches of long grass.
- seasonal signs occur in summer and autumn:
  - resolution of most signs in winter and spring.

- incidence may be reduced by regular use of avermectin anthelmintics.

Clinical presentation

- granulating sores predominately:
  - facial skin below the medial canthus of eye.
  - prepuce and urethra.
- rapid expansion of non-healing granulating wounds:
  - possible stringy serosanguineous exudate.
- multiple areas can be affected (Figs. 4.58, 4.59).
- infestation of the urethral process can lead to:
  - sheath haemorrhage.
  - blood-stained urine, dysuria or pollakiuria.
  - possible bleeding during mating by stallions (reduced fertility).
- caseous sulphur-like granules ('kunkers') occur in some sores:
  - particularly evident in medial canthus and conjunctival forms.
- wound infestation at other sites can produce large tumour-like growths ('summer sores'):
  - common in preputial region or on penis.
- variable pruritus but when severe can cause skin damage and expanding lesions:
  - rubbing of eyes and biting at lesions occurs frequently.

FIG. 4.58 This 9-year-old grey gelding had a history of repeated summer-related bilateral dermatitis on the face. Habronemisis was diagnosed.

FIG. 4.59 This gelding developed a rapidly progressive granulomatous lesion in several places within the preputial skin. A diagnosis of habronemiasis was confirmed by biopsy and PCR.

Diagnosis

- history, clinical signs, presence of sulphur granules and cytological and histopathological identification of larvae:
  - larvae in washings taken from the area with a soft toothbrush.
- eggs are not usually identified in faeces.

Management

- Ivermectin (0.2–0.3 mg/kg) p/o is treatment of choice:
  - occasionally a second dose is required 3–4 weeks after the initial dose.
- eye lesions may require local application of corticosteroid drops plus a 5% mixture of injectable ivermectin in artificial tears solution.
- hypersensitivity reactions are treated by:
  - oral prednisolone (1 mg/kg q24 h for 10 days).
  - topical betamethasone skin cream mixed with 10% oral ivermectin wormer.
- surgical removal or debulking of larger granuloma lesion in severe cases and particularly within the sheath:
  - complete resolution usually requires topical treatments.
- attention to fly control and wound management minimises recurrence.
- average reported recovery time is approximately 23 days.

Prognosis

- good, but all aspects of control as well as treatment must be followed.
- cases usually affected annually with progressively deteriorating clinical signs:

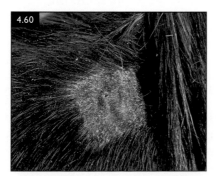

FIG. 4.60 *Hypoderma bovis* lesion in the skin. Extrusion of L3 larvae from a nodule or cyst with a breathing 'pore'. (Photo courtesy A Waddell.)

- prophylactic/pre-emptive management strongly advised in all cases.

Hypodermiasis

Definition/overview

- caused by *Hypoderma bovis* (warble) fly larvae.
- migratory life cycle with tissue invasion before undergoing final development in skin.
- very rare in northern hemisphere as bovine form of the disease largely eliminated by strict legislative control.

Aetiology/pathophysiology

- adult flies attach their eggs to horse's hair in a row.
- larvae hatch and penetrate underlying tissues to undertake a variable tissue migratory phase:
  - L2 larvae migrate to withers and dorsal back.
  - cause localised painful subcutaneous swellings.

Clinical presentation

- firm, often mildly painful, inflamed nodule in dorsal midline, especially withers (occasionally elsewhere).
- development of breathing 'pore' from which the mature larva may or may not emerge (Fig. 4.60).
- pruritus may be present during development of larvae but usually site is hot and mildly painful.

Differential diagnosis

- epidermoid and dermoid cyst.
- infectious granuloma
- Onchocerca infection.
- neoplasia of various types including sarcoid and melanoma.
- equine eosinophilic granuloma.

Diagnosis

- difficult to confirm in early stages but ultrasonography is helpful.

Management

- surgical removal of entire nodule.

- routine oral deworming with ivermectin or moxidectin usually prevents larval migration and growth:
  - consequent anaphylactic reactions to dead larvae have been reported.

Prognosis

- good even if the lesion is allowed to mature.
- scarring if secondary infection occurs or the parasite dies before emerging.

Insect bites or stings

Definition/overview

- many types of insects that produce varied bite distribution on horses.
- complete clinical history is necessary to eliminate many likely agents.
- adult female of most free-living Diptera only requires one blood feed to lay eggs, and therefore difficult to control.

Clinical presentation

- wide variety of stings and bites occur in horses:
  - biting flies, bees, wasps, hornets and spiders etc. can be involved in different geographical and climatic conditions.
- most bites and stings show a central focus surrounded by a small circular oedematous plaque:
  - the larger the fly, the larger the bite puncture and area of oedema.
- irritability (e.g. stamping feet, restless movements, rubbing and galloping around the paddock) due to pain/annoyance caused by biting insects.
- can be intensely pruritic.

Diagnosis

- biting insects usually more prevalent in warm months/after rain.
- presence of flies, midges, bee swarm or wasps, or contact with insect nests.

Management

- removal of the causal agent or avoidance of contact opportunities.

Prognosis

- single or few bites, invariably good:

- in some countries, the insects are more aggressive and cause significant consequences:
  - ◆ massive anaphylactic responses resulting in death with *Simulium* spp. fly attack in USA.
  - ◆ swarm attacks by bees and wasps.
  - ◆ spider bites in Australia can be very serious.
- vector-transmitted diseases are important and control of many of these is reliant upon control of the vectors responsible for their transmission.

Myiasis

Definition/overview

- infestation of tissue by live larvae caused by several fly species including:
  - *Lucilia* spp. ('Greenbottle fly').
  - *Calliphora* spp. ('Bluebottle fly').
  - *Phormia* spp. ('Blackbottle').
  - *Cochliomyia* spp. flies.
  - *Calliphora* spp. (blow fly strike).
- some species feed only on necrotic tissue.
- others such as *Wohlfahrtia* spp. and *Chryosomya* spp. are more aggressive and destructive for live tissue.
- medical-quality *Lucilia sericata* larvae are used therapeutically in the management of wounds or hoof canker.

Aetiology/pathophysiology

- rapid seasonally restricted infestation of:
  - neglected wounds   ○ plaster casts.
  - dirty bandages.
  - necrotic tissue including neoplasms such as sarcoid or carcinoma.
- serious disease with destructive infestation of fresh clean and healthy tissue injuries:
  - screw-worm larvae *Cochliomyia hominivorax* and *Callitroga macellaria* in North, Central and South America.
  - *Chrysomya bezziana* in Africa and Asia.

Clinical presentation

- irritation resembling pruritus usually noticed early in the development:
  - chewing at wounds and bandages.
- characteristic malodour and discharge.
- sometimes only the breathing pore end of larvae is visible within wounds (Fig. 4.61).

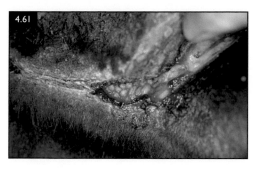

FIG. 4.61 Myiasis due to *Calliphora* spp. larvae in a wound.

Differential diagnosis

* necrotic infection of wounds.

Diagnosis

* examination of the wound.

Management

* wound hygiene is essential.
* larvae relatively easily destroyed by washes or ointments containing ivermectin.

Prognosis

* good, provided debridement and follow-up prevention carried out properly.
* repeated infestations associated with a breakdown in management.

IMMUNE-MEDIATED AND ALLERGIC CONDITIONS

Insect bite hypersensitivity

Definition/overview

* 'sweet itch' or 'Queensland itch' is the most common skin allergy in horses worldwide.
* no sex, colour or skin colour predilection.
* rare before 3 years of age but progression over sequential years is common.

Aetiology/pathophysiology

* intense regional pruritus from types 1 (immediate and late/delayed) and 4 (delayed) hypersensitivity reactions to specific salivary proteins mainly, but not exclusively, from *Culicoides* spp. midges:
  ○ other biting flies (*Simulium* spp., *Stomoxys calcitrans* and possibly *Haematobia irritans*) also sometimes responsible.
  ○ geographical distribution of many *Culicoides* spp. midges varies:
    ♦ cause can be specific to midges (or other insects with similar salivary antigens) occurring in restricted areas.
* familial and genetic basis/susceptibility or predisposition in Icelandic, Shire, Welsh and Arabian ponies/horses:
  ○ exposure to appropriate allergen is essential.

Clinical presentation

* rubbing of tail, neck/base of mane, head and back almost always involved (Figs. 4.62, 4.63).
* geographical location and prevalent midge species may affect incidence of lesions elsewhere including on ventral midline or on face and ears.
* exfoliation, plasma exudation, patchy alopecia, and crusting can occur with rubbing.
* thickened skin rugae can develop on the withers, neck, tail head and ventral midline with chronicity (Fig. 4.64).

FIG. 4.62 Early case of *Culicoides* spp. hypersensitivity with papules, alopecia, slight scurf and the mane extensively rubbed out, due to severe pruritus. (Reprinted from Pascoe RR (1990) *Colour Atlas of Equine Dermatology*, Wolfe, with permission.)

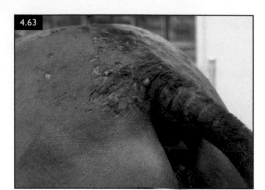

FIG. 4.63 *Culicoides* spp. can affect rugged horses without tail protectors, resulting in rubbed-out tail hairs. These lesions had been present for over 3 weeks.

- chronic hair loss due to mechanical irritation from rubbing.
- possible loss of weight due to constant irritation.
- generally, affected horses itch more in the early evening (dusk) and morning (dawn).

Diagnosis
- clinical signs and seasonal recurrence are highly suggestive:
  - other ectoparasitic causes should be excluded.
  - **hypersensitivity can be associated with Tabanidae (horse flies), *Stomoxys* spp. (stable flies), *Haematobia* spp. (buffalo flies) or a mixture.**
- improvement on restriction to indoor management or rugging etc. is highly suggestive.
- biopsy shows non-specific dermatitis with mild to severe eosinophilic folliculitis:
  - chronic lesions often unrewarding.
- intradermal skin testing using aqueous whole insect or recombinant salivary antigens:
  - reported to give reliable positive results.
  - specific species antigens relating to prevalent *Culicoides* in geographic area give best response.
- **further diagnostic tests for insect hypersensitivity include:**
  - stringent ectoparasite challenge control trial using 2% permethrin products applied up to every other day.
  - intradermal skin testing to identify allergen sensitivities to other insects

FIG. 4.64 Old lesions due to *Culicoides* spp. still show severe permanent damage to the skin and mane hair, and lichenification of the skin, in mid-winter.

including stable flies, mosquitoes, deer flies, horse flies and black flies.
  - isolation of *Culicoides* salivary gland extracts has identified cases with 100% accuracy:
    - *in vitro* testing should improve with the development of assays using specific recombinant salivary gland extracts.
- serological testing is generally unreliable.

Management
- **multimodal approach provides the best outcome in highly susceptible horses.**
- avoidance strategies are the best management approach for any allergic condition:
  - complete avoidance is not usually possible.
  - stabling is a good measure to take immediately.
  - turnout restricted to times when the midges are not active:
    - avoid early morning and evening.
    - stable between 16:00 and 08:00.
  - rugging/blanketing with sheets and hoods may prevent serious skin damage.
  - move the horse to an area where midges are less prevalent, such as:
    - windy hillside.
    - away from river courses and woodland environments.
  - insect-proofing stables using high-velocity fans is beneficial (difficult to achieve).
- removing all available 'rubbing' sites such as fencing and trees etc. by surrounding

the horse with an electric fence is not considered humane and should not be used if possible.

- insect repellents are widely used with limited results:
  ○ can help alongside other measures.
  ○ dietary garlic said to improve the condition by repelling insects (not established).
- some cases respond (partially) to antihistamine administration (Table 4.1).

TABLE 4.1 Antihistamines

ANTIHISTAMINE	DOSE	FREQUENCY
Cetirizine	0.2–0.4 mg/kg	q12 h
Hydroxyzine hydrochloride or pamoate	0.5–1.0 mg/kg	q8 h
Doxepin hydrochloride	0.5–0.75 mg/kg	q12 h
Chlorpheniramine	0.25 mg/kg	q12 h

- daily oral treatment with prednisolone or dexamethasone during the high-risk season may reduce the severity of the pruritus:
  ○ secondary effects from prolonged administration are rare but should be constantly reviewed.
  ○ longer-lasting effects in seriously affected horses can be gained by administration of methylprednisolone (0.2 mg/kg i/m at 3–4-week intervals):
    ♦ **prolonged usage is not advisable.**
- treatment of unrugged paddock horses is extremely difficult:
  ○ concentrated permethrins applied to dorsum of horse may give some relief:
    ♦ **paraesthesia in response to high doses of permethrins.**
    ♦ **pre-application of concentrated topical vitamin E at treatment sites (1,000–2,000 IU).**
- other strategies to decrease the exposure to *Culicoides* spp. midges and flies include:
  ○ malathion spray 500 g/l:
    ♦ 125 ml/10 litres of water initially as a spray and repeat in 7 days.

- wipe-on chemical such as Coopers Fly Repellent Plus™ (permethrin and citronella) or Deosan Deosect™ 5% (cypermethrin).
- shampoos containing insecticides (usually mixture of pyrethrins and other synthetics) but need frequent repetition.
- weekly body sprays with synthetic pyrethroids, such as:
    ♦ 2–3.6% permethrin and 5% cypermethrin:
      – greater than 80% midge mortality after 7 days.
      – residual activity of 50% at day 35.
    ♦ Cyfluthrin (fourth-generation pyrethroid):
      – 10 times the potency and extended residual activity (30 days).
    ♦ topical high-concentration 44% permethrin-containing canine flea products:
      – used off-label every 2–4 weeks with success.
- applied daily to affected sites of skin damage:
  ○ topical steroids including 0.1% mometasone ointment combined with an antibiotic and antifungal agent.
  ○ or 0.584 mg/ml esterified hydrocortisone aceponate spray (Cortavance™, Virbac).
- long-term supplementation with polyunsaturated n-3 (fish or algae oil) and n-6 (evening primrose oil) fatty acids (PUFAs):
  ○ may stabilise mast cells and reduce pro-inflammatory mediators.
- Phosphodiesterase inhibitors such as pentoxifylline:
  ○ 8–10 mg/kg p/o q8–12 h.
  ○ anti-inflammatory alternative to steroids with minimal side effects.
- immunotherapy using repeated injections of recombinant *Culicoides* allergens into submandibular lymph nodes has had some benefit.

Prognosis

- cases seldom resolve completely, and lifelong strategies may be necessary.

- early-life exposure to *Culicoides* spp. midges within the first 6 months of life is helpful in reducing the severity of the disease in susceptible animals.
- aggressive avoidance strategies are more effective than adjunctive treatments.
- intractable cases that cannot be managed in their environment may require euthanasia.

Atopic dermatitis (atopy)

Definition/overview

- genetic/heritable allergy that does not require prior exposure to inhaled, ingested or contacted allergen:
  - possibly more common in Arabian, Thoroughbred, Quarter Horse and Warmblood breeds and their crosses.
  - no clear sex predisposition.
- relatively common but underdiagnosed, chronically relapsing pruritic skin disease.

Aetiology/pathophysiology

- unclear cause and pathogenesis.
- exposure to defined allergens trigger immediate (type 1) hypersensitivity with mast cell degranulation producing pruritus and/or urticaria.
- allergens include inhaled dust or protein particles and ingested or contacted allergens.
- respiratory signs may also occur.

Clinical presentation

- chronic recurring seasonal or non-seasonal pruritus, with or without urticaria:
  - starts at a relatively young age (generally over 2–3 years old).
  - progressive signs year on year.
  - pruritus and urticaria mainly on the body and neck:
    - occasionally on the limb and head.
  - symmetrical shifting pruritus is common:
    - marked in focal areas with minimal skin damage.
  - occasionally moderate to severe skin damage with lichenification over time:

- sometimes mane and tail hair loss.
- possibly associated with airway obstructions and equine asthma syndromes.

Differential diagnosis

- other allergic conditions including insect bite hypersensitivity (may occur concurrently).
- insect-related pruritus, food allergies and contact hypersensitivity.

Diagnosis

- confirmation of diagnosis is challenging:
  - intradermal skin tests and patch testing usually identify multiple positive allergens:
    - regarded as 'difficult to interpret'.
  - serum IgE testing is presently unreliable.
- skin biopsy of pruritic sites usually unrewarding or non-specific.

Management

- primary objective is to reduce the pruritus and consequent skin damage.
- rugs and blankets often reduce the itch but when removed the patient will show an aggressive itch.
- elimination of an identified responsible allergen is helpful but difficult:
  - environmental, diet and management changes can help but are usually unrewarding.
- steroid and antihistaminic drugs are often helpful but usually only short term.
- antigen-specific immunotherapy can be helpful where the allergen is known:
  - 50% of cases show a 50% reduction is clinical signs after 'vaccination'.

Prognosis

- seldom life-threatening or resolved.
- best viewed as incurable but tolerable in most cases, with owner advice and support essential.

Urticaria

Definition/overview

- specific skin lesion rather than a specific disease entity:

- clinical manifestations vary from a minor transitory nature to major, systemic, life-endangering problems.
- many different aetiologies and pathogeneses.

Aetiology/pathophysiology

- degranulation of mast cells and basophils is the basic pathogenesis:
  - liberates chemical mediators leading to increased vascular permeability.
  - wheal formation.
- immunological/allergic:
  - hypersensitivity develops when an antigen/allergen is introduced:
    - local contact or injection (insect allergens, drugs).
    - ingestion (chemical, feeds).
    - inhalation and transepidermal absorption (pollen, dust, chemicals, moulds).
- physical urticaria:
  - non-immunological pathogenesis.
- Dermatographism: wheal developing from a blunt scratch on skin.
- urticaria due to cold, heat or light.
- exercise-induced urticaria.

Clinical presentation

- acute to peracute onset with signs developing within minutes to a few hours.
- oedematous lesion of the skin or mucous membrane develops, called a wheal:
  - flat-topped papule/nodule with steep walled sides, which pits on pressure:
    - some have slightly depressed centres.
    - vary in size and shape:
      - conventional: 2–3 mm up to 3–5 mm (Fig. 4.65).
      - papular:
        ◊ multiple small, uniform, 3–6 mm diameter wheals (e.g. insect bites).
      - giant:
        ◊ single or coalesced multiple wheals up to 20–30 cm diameter.
      - gyrate annular: serpiginous or arciform lesions.
      - linear:
        ◊ bilaterally symmetrical, parallel bands of urticaria over the trunk.

- lesion may present differently depending on the location of the oedema:
  - whether in upper layers of the dermis or subcutaneous tissue.
- 'Oozing' urticaria:
  - dermal oedema is severe.
  - serum may ooze from the skin surface, forming crusted lesions.
  - **distinguish from erosive/ulcerative process, pyoderma or pemphigus foliaceus:**
    - lesion still pits on pressure.
- Angioedema (angioneurotic oedema):
  - subcutaneous form which tends to be more diffuse:
    - spread in the subcutis due to lack of resistance.
  - usually involves the head and extremities.
  - more indicative of a systemic and serious disease than urticaria.
  - pruritus may or may not be present.

Differential diagnosis

- insect (*Stomoxys* spp. and *Culicoides* spp.) or mosquito bites.
- bee and wasp stings.
- cellulitis and possibly vasculitis.
- purpura haemorrhagica.
- erythema multiforme • haematoma.
- lymphangitis • alopecia areata.
- pemphigus.
- *Trichophyton* spp. infestation.

Diagnosis

- clinical signs and history.

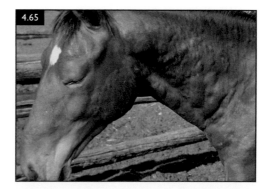

FIG. 4.65 Urticarial lesions of 2–5 mm diameter spread all over the body due to a change of feed, which disappeared in 48 hours.

- skin biopsy helpful in eliminating other potential differential diagnoses.
- 'Cold' urticaria test:
  - ice cube applied to the skin leads to oedema within 15 minutes indicates a positive response.
- environmental allergen testing using intradermal or serological allergy skin testing.
- food allergy is reported to be uncommon:
  - many cases are, however, significantly improved by avoiding cereal foods.
  - only confirmed by elimination diets followed by challenge with suspected feed.

Management

- best treatment is avoidance of the allergens – assuming they can be identified.
- cases due to vasoactive amines in cereal foods:
  - reducing cereal content of diet may be effective or partially effective.
- short course of systemic corticosteroids:
  - repeated if signs recur.
- persistent urticaria for 8 weeks or more requires further investigation:
  - intradermal skin or serological IgE testing.
  - identify allergens for inclusion in an allergen-specific immunotherapy (ASIT) treatment set.
- long-term use of corticosteroids and/or antihistamines required in some cases:
  - oral administration of dexamethasone at lowest possible dose on alternate days.

- oral antihistamine with cetirizine hydrochloride at 0.2–0.4 mg/kg q12–24 h:
  - at least 2 weeks to allow full effects.
- Doxepin (0.5–0.75 mg/kg q12 h) antidepressant with antihistaminic activity:
  - sometimes used successfully for patients that do not respond to cetirizine.

Equine pemphigus foliaceus

Definition/overview

- autoimmune disease characterised by:
  - variable areas of alopecia and exfoliative dermatitis.
  - pustular disease with the formation of heavy crusting (Figs. 4.66, 4.67).

Aetiology/pathophysiology

- unknown what triggers the autoimmune response:
  - many potential factors have been identified:
    - co-morbidity.
    - environmental changes.

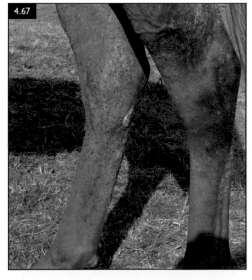

FIG. 4.67 Equine pemphigus foliaceus lesion on the thigh and metatarsal region. These can closely resemble bacterial infections such as Dermatophilosis.

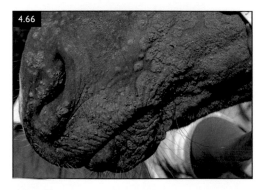

FIG. 4.66 Equine pemphigus foliaceus with crusts and scale on the muzzle and face.

- ♦ dietary challenges.
- ♦ pregnancy   ♦ foaling.
- ♦ drug administration.
- ♦ vaccination.
- autoimmune attack targets the desmosomes in the upper layers of the epidermis:
  - ○ destruction causes separation of keratinocytes resulting in epidermal clefting detectable histologically.
  - ○ exfoliation of heavy plates of loosely attached keratinised epidermis:
    - ♦ characteristic scaly nature.
    - ♦ formation of subcorneal pustules that frequently rupture leaving epidermal collarettes, crusts and scale as visible clinical lesions.

Clinical presentation

- early cases show transient vesicles, erosion, epidermal collarettes, crusting, scaling, fever, depression, weight loss, coronary band lesions, variable pruritus and pain.
- advanced cases show severe crusting and scaling, alopecia and poor appetite.
- pitting oedema of the limbs, the ventral abdomen and sheath is also common.

Differential diagnosis

- dermatophilosis   • dermatophytosis.
- onchocerciasis.
- *Culicoides* dermatitis.
- equine viral papular dermatitis.
- equine sarcoidosis.
- epitheliogenesis imperfecta.
- junctional mechanobullous disease.
- coronary band dystrophy   • seborrhoea.
- equine granulomatous enteritis.
- generalised skin eruptions of unknown aetiology.
- *Malassezia* spp. (yeast) dermatitis.

Diagnosis

- history and clinical signs are strongly suggestive in most cases.
- rapid diagnosis sometimes achieved by staining direct impression smears taken from crusts, intact vesicles or pustules:
  - ○ reveal acantholytic cells among extensive neutrophils.
- multiple skin biopsies from multiple sites are diagnostic:

- ○ important to include the overlying crust.
- ○ clipping, washing and wiping avoided as these may remove the crust.
- ○ histopathology is pathognomonic.
- ○ photographic images including biopsy sites are helpful to the pathologist.
- direct immunofluorescence testing or immunohistochemistry also useful.

Management

- immunosuppressive doses of corticosteroids are indicated:
  - ○ prednisolone 2–4 mg/kg q12 h for 7–14 days.
  - ○ attempts made to establish a minimal effective once-daily dose.
  - ○ gradually then work towards a minimum effective alternate daily dose which can be maintained for extended period (up to years).
- Azathioprine (3–5 mg/kg q24 h) can be used (corticosteroid-sparing medication).
- Pentoxifylline (8–10 mg/kg q12 h) may also provide anti-inflammatory and steroid-sparing effects.
- treatment usually needs to be maintained for long periods and possibly even for life:
  - ○ rarely, usually in younger horses, medication can be slowly tapered and discontinued.

Prognosis

- always very guarded.
- long periods of remission and control may be followed by severe disease requiring heavy medication.
- progressive deterioration despite increasing medication is a poor prognostic sign.
- young horses (<1 year of age) may have milder clinical signs and respond more favourably.

Equine pemphigus vulgaris and bullous pemphigoid

Definition/overview

- extremely rare vesicobullous ulcerative disease.
- most cases affect the oral cavity (Fig. 4.68), mucocutaneous junction or skin, or a combination of all three.

FIG. 4.68 Bullous pemphigoid erosion in the mouth. (Photo courtesy DW Scott.)

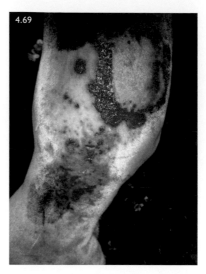

FIG. 4.69 12-year-old Irish Sport Horse gelding developed 'pastern dermatitis' on the lateral aspect of its left hind leg ascribed by the owner to 'mud fever'. The limb showed intermittent swelling which reduced on exercise. Careful examination and clipping revealed the pathognomonic lesion distribution and appearance of pastern and cannon leukocytoclastic vasculitis which was confirmed histologically. Other limbs were not affected.

Pastern and cannon leukocytoclastic vasculitis (PCLCV) (cutaneous small-vessel vasculitis)*

Definition/overview

- increasingly prevalent disease affecting non-pigmented pastern and cannon regions:
  - may only affect one limb even when other limbs are non-pigmented.
  - lateral aspects of non-pigmented hindlimbs are most often affected.
- affects individual mature horses without breed, sex or colour predilection.

Aetiology/pathophysiology

- cutaneous, small-vessel vasculitis of the dermal capillaries and venules.
- idiopathic or associated with infection, neoplasms, autoimmune disorders and drugs.
- possibly mediated by immune-complex deposition relating to buttercup plant (*Ranunculus* spp.) exposure.
- may be triggered by drug reactions, food antigens and/or environmental allergens.
- exacerbated by exposure to ultraviolet light:
  - most often occurs in summer in regions with plentiful sunlight.
  - exposure to snow can also induce an extreme response.
  - non-pigmented skin involvement.
  - photo-exacerbated disorder and not a true photosensitisation.

Clinical presentation

- pastern and cannon regions of non-pigmented limbs most commonly affected:
  - hindlimbs affected more than front limbs.
  - lateral more common than medial.
  - usually little or no involvement of dorsal/plantar/palmar aspects of the limbs unless secondary infection develops.
- early lesions clearly demarcated marginal erythema and crusting with variable changes centrally (Fig. 4.69):
  - result of red cell extravasation in acute stage of condition.
  - often only visible after clipping the hair away – otherwise a line or wider area of crusting may be all that shows at first.

---

* 'Leukocytoclastic' refers to the damage caused by nuclear debris from infiltrating neutrophils in and around the vessels.

- later lesions show erosions and ulceration with some oedema of the affected limb.
- crusting often difficult to remove and attempts can be painful.
- no pruritus.
- some cases have a wider more diffuse inflammation (Fig. 4.70); more rarely, a multinodular form showing the same histological response can develop.

Differential diagnosis

- photosensitisation, especially due to primary plant ingestion or hepatic failure.
- bacterial dermatitis, particularly dermatophilosis due to *D. congolensis* infection.

Diagnosis

- clinical appearance including involvement of non-pigmented pastern and cannon.
- biopsy is definitive in most cases.

Management

- Table 4.2:
  - time-consuming but avoids using systemic medication or topical creams/lotions in most cases.
  - do not use iodine-based soaps.
- stable out of sunlight (or snow during daylight hours) to avoid exposure to UV light and until 4 weeks after normal skin is achieved.

- exercise maintained to reduce limb oedema and maintain a healthy skin microbiome, but without sunlight exposure:
  - indoors and pre-dawn/post-sunset are acceptable.
- management strategies for up to 3 months.
- gradual reintroduction to sunlight to limit exacerbation or recurrence.
- medications usually have a mild and transient benefit:
  - reserved for non-responsive cases or where UV light exposure is unavoidable.
  - usually do not solve the problem and may prolong it.
  - high-dose glucocorticoids:
    - prednisolone (1 mg/kg q12 h) or dexamethasone (0.1 mg/kg q24 h) for 2 weeks, tapering over 4–12 weeks.
  - Pentoxifylline (8–10 mg/kg q12 h).
  - topical corticosteroids and other creams/lotions should be avoided.
  - antibiotics are not helpful or required unless secondary infection is present.
- boots, bandages and stockings etc. are not helpful.
- eliminate exposure to potential causes including any dietary or other supplements.

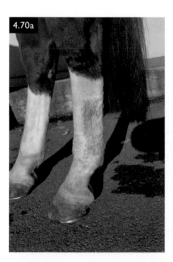  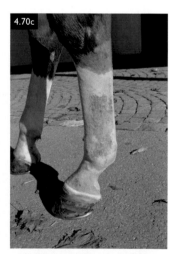

FIG. 4.70 More diffuse form of PCLCV with the true severity masked by the hair **(a and b)** and only revealed after the hair is removed **(c)**. Medial and lateral aspect of the left hind were affected but the right hind and both front legs (also white) showed no pathology.

TABLE 4.2 'The hot wash system'

PROCEDURE	DURATION	COMMENTS
clip all hair with surgical clippers	open	wider than the condition. usually entire limb distal to carpus/tarsus
crusting or pain preclude clipping. Apply wound hydrogel under a cling film wrap	2–3 hours	crusting must be removed
apply neat chlorhexidine surgical scrub over whole clipped area but particularly affected regions	15 minutes	leave on leg
Using warm to hot water (40ºC), shampoo region vigorously	10 minutes	**do not use cold water**
leave on after shampoo completed	15–20 minutes	keep the horse controlled
pour clean warm water (40ºC) over the site until it runs clean	10 minutes	**do not touch/handle skin.** **do not use cold water**
dry skin by blotting only	5 minutes	**do not rub dry.** only use paper towel. dispose towel after each blot. limb must be dry before it cools
severe inflammation is present: spray once only with hydrocortisone aceponate	n/a	avoid if possible
leave site open to air but no exposure to sunshine	4–7 weeks	dressings and bandages are unhelpful and can make it worse. avoiding UV exposure is critical

Prognosis

- good as recovered cases seldom recur.
- recovery requires meticulous local skin care with strict discipline in avoiding sunlight exposure.
- some cases are more intractable, serious and have a poor prognosis:
  - these may benefit from a case review to establish any underlying causes and other differential diagnoses.

Vasculitis

Definition/overview

- uncommon disorder characterised by purpura, oedema, necrosis and ulceration.
- particularly affects the distal limbs and oral mucosa.

Aetiology/pathophysiology

- possibly mediated by immune complex deposition:
  - particularly post-strangles or other respiratory infections.
- circulating antigens induce antibody formation:
  - triggered by factors such as medications, infections, food antigens, environmental allergens or neoplasia.
- antibodies bind the circulating antigen and create immune complexes:
  - deposited within vessels, activating complement and inducing inflammatory mediators.
- inflammatory mediators, adhesion molecules and local factors affect endothelial cells.
- exacerbation of lesions by photo-aggravation is common.

Clinical presentation

- common sites are the coronet, pastern, fetlock (Fig. 4.71), lips and periorbital tissues.
- usually not restricted to white skin.

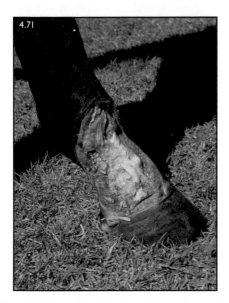

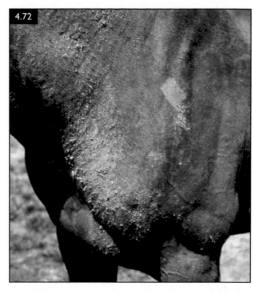

FIG. 4.71 Vasculitis following *S. equi* vaccination during a severe outbreak of strangles on a Standardbred farm. This yearling had lesions on all four pasterns and fetlocks.

FIG. 4.72 Equine sarcoidosis in a 7-year-old gelding with chronic wasting disease, marked exfoliative dermatitis and heavy scurf on the shoulder and neck. (Photo courtesy JR Vasey.)

- oedema, erythema, necrosis and ulceration can occur.
- pyrexia, depression, anorexia and weight loss may also be present.
- no pruritus or pain, except in early crusts.

Diagnosis

- history and clinical appearance suggestive.
- biopsy from a fresh lesion in the first 24 hours provides confirmation.

Management

- any underlying disease should be treated.
- early cases may respond to corticosteroids:
  - oral prednisolone (1–2 mg/kg) twice daily until regression occurs.
  - then reduced to lowest possible alternate-morning dose.
  - Pentoxifylline (8–10 mg/kg twice daily) – useful adjunctive and steroid-sparing agent.
- minimise direct sunlight exposure.

Prognosis

- guarded but dependent on the cause.

- cases identified and treated early respond better in the long term.

Equine sarcoidosis

Definition/overview

- rare systemic granulomatous disease that presents as either:
  - localised cutaneous form with exfoliative to nodular dermatitis (Fig. 4.72).
  - generalised granulomatous disease affecting most internal organs:
    - results in systemic signs including weight loss, poor appetite, persistent low-grade fever and exercise intolerance.
- many breeds are affected, with mares having a greater predisposition,
- onset of clinical signs is typically at 3 years of age or greater.
- prognosis for the generalised granulomatous disease is generally poor.
- localised cutaneous form has a favourable prognosis but typically requires lifelong treatment.

Equine cutaneous lupus erythematosus (CLE)

Definition/overview

- rare, incompletely defined, clinicopathological entity found in the horse.
- differs from the classic forms seen in humans and dogs (Fig. 4.73).

Aetiology/pathophysiology

- autoimmune disease.
- pathogenesis is multifactorial:
  - may be localised to the skin only.
  - severe cases, may induce multisystemic signs of autoimmune attack:
    - polyarthritis, vasculitis and dermatitis.

Clinical presentation

- sharp demarcation between pigmented and depigmented areas.
- depigmentation is seen around:
  - eyes, lips, nostrils, vulva, anal ring and prepuce (Figs. 4.74, 4.75).
  - loss of pigment can be gradual or rapid.
- erythema and scaling with alopecia:
  - skin of long-standing cases looks like wrinkled leather.
- ANA testing is negative.
- Systemic lupus erythematosus (SLE) cases may also show:
  - organ dysfunction, weight loss and fever.
  - often have a positive ANA titre.

Diagnosis

- skin biopsy:
  - specimens in formalin for histopathology and immunohistochemistry evaluation.
  - place in Michel's medium if submitted for immunofluorescence studies.
- serum for an ANA test and quantification of IgM and/or IgG may also be submitted.

FIG. 4.73 Equine cutaneous lupus erythematosus.

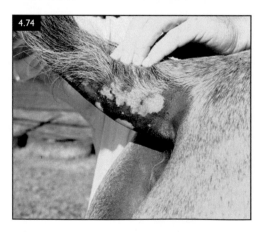

FIG. 4.74 Equine cutaneous lupus erythematosus was diagnosed in this grey 12-year-old Arabian stallion with depigmentation in areas of the neck, flank and tail. Depigmentation of skin under the tail was very clear.

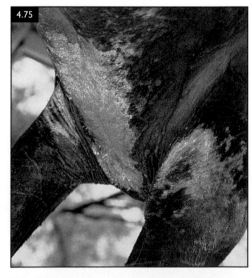

FIG. 4.75 Systemic lupus erythematosus in a 16-year-old stock horse that had severe exfoliation of the body, neck and head, with marked depigmentation of the elbow area. Post-mortem examination revealed a chronic pleuritis and abscessation on the left lateral chest wall.

FIG. 4.76 Equine alopecia areata in a 13-year-old grey Arabian mare with patchy alopecia of 5 years' duration. Biopsy revealed only very small areas of lymphocytic bulbitis in hair follicles.

Differential diagnosis
- other depigmentation diseases:
  - Arabian fading syndrome.
  - Equine granulomatous enteritis.
  - Leucoderma.

Management
- localised lesions treated with topical glucocorticoids or topical tacrolimus 0.1%.
- SLE cases require:
  - systemic tapering immunosuppressive doses of glucocorticoids.
  - adjunctive immunomodulatory medications:
    - azathioprine or pentoxifylline.

Prognosis
- fair for CLE and poor for SLE.

Equine alopecia areata

Definition/overview
- cell-mediated 'autoimmune' skin disease.
- uncommon but more prevalent in the horse than other domestic species.

Aetiology/pathophysiology
- T lymphocyte and dendritic immunological attack on anagen hair follicles.
- early diseased hair follicles may show defective keratinisation.

- inflammation may be minimal in chronic cases and reveal small follicles lacking hair.
- possible hereditary factor as 20% of reported cases are familial.

Clinical presentation
- one or more reasonably circumscribed areas of partial or complete alopecia:
  - variable areas of facial and body hair, and the mane and tail (Fig. 4.76).
  - areas can coalesce to produce extensive alopecia.
- relatively slow onset in most cases.
- non-pruritic and no visible signs of inflammation.
- defective hoof growth may occur.
- partial or complete spontaneous remission occurs rarely.

Differential diagnosis
- Dermatophytosis
- dermatophilosis.
- occult sarcoid.
- anagen effluvium (defluxion).
- telogen effluvium (defluxion).
- follicular dysplasia.

Diagnosis
- skin biopsies of multiple stages of alopecia:
  - usually confirm the diagnosis in early stages.
  - chronic lesions may not show pathognomonic changes.

Management
- treatment is generally not recommended.
- early cases may respond to immunosuppressive drugs, but results are usually poor.

Prognosis
- condition is purely cosmetic.

Erythema multiforme

Definition/overview
- rare, acute, self-limiting, urticarial maculopapular or vesicobullous dermatosis.
- most cases are classified as idiopathic.

- triggering factors include drugs, infections (especially herpesvirus) and tumours (especially lymphoreticular neoplasms).
- characteristic 'doughnut-like' or serpiginous urticarial skin lesions develop rapidly from the initial urticaria and plaques (Fig. 4.77).
- advanced cases can present with ulcerative lesions and become systemically ill (febrile, anorexic and depressed).

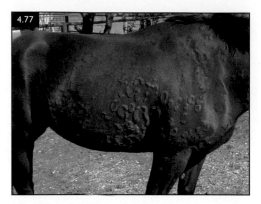

FIG. 4.77 Erythema multiforme. Characteristic doughnut-like lesions of urticaria-like focal skin swelling.

NUTRITIONAL AND TOXIC CONDITIONS

Photosensitisation dermatitis
(see also page 3, Chapter 1)

Definition/overview
- caused by UV radiation and facilitated by lack of pigment and hair.

Aetiology/pathophysiology
- two categories:
  - excessive sunlight exposure (sunburn).
  - normal sunlight exposure with an unexpected outcome (photosensitisation).
- photosensitisation is related to three factors:
  - presence of a photodynamic agent within the skin.
  - exposure to sunlight or certain wavelengths of UV light.
  - cutaneous absorption of this UV light.
- photosensitisation may be:
  - systemic condition due to primary ingestion of a photodynamic agent from the digestive tract with transfer into the skin via the circulation:
    - St John's wort (*Hypericum perforatum*) or other plant species.
  - hepatogenous:
    - phylloerythrin accumulating in tissues (photodynamic agent).

Clinical presentation
- usually restricted to light skin or hairless areas:
  - severe cases may extend into dark-skinned or haired areas.
- commonly affects the lips, face, eyelids, perineum and coronary band region.
- conjunctivitis, oedema, erythema, pruritus, pain, oozing, necrosis and sloughing of skin.

Differential diagnosis
- Dermatophilosis
- greasy heel.
- dermatophytosis.

Diagnosis
- liver function tests should be performed.
- history of pasture grazing, treatments and diet should be investigated:
  - plants can be tested for pyrrolizidine alkaloids.
- lesions due to pasture plants, sprays or drugs:
  - may be localised to the lip and lower limb.
- non-hepatogenous cases:
  - multiple horses grazing in the same paddocks may be affected.
- biopsy reveals non-specific histopathological changes.

Management

- protect from direct sunlight by stabling, hoods, rugs or other means.
- emollient creams can be applied.
- sources of photodynamic agents should be eliminated.
- therapy for hepatic disease, if present, should be provided.

- glucocorticoids, pentoxifylline and NSAIDs may assist in reducing inflammation.

Prognosis

- most affected animals recover, but severe cases may have residual skin scarring.
- cases associated with liver disease carry a poor prognosis.

ENDOCRINOLOGICAL DISEASE

Pituitary pars intermedia dysfunction (PPID) (equine Cushing's disease)
(see page 25, Chapter 2)

IDIOPATHIC CONDITIONS

Vitiligo (Arabian fading syndrome)

Definition/overview

- most common in Arabians.
- characterised by annular areas of macular depigmentation of the muzzle, lips, around the eyes and occasionally around the perineum, sheath and hooves.
- occurs at any age but usually found in horses over 4 years of age.
- occasional body patches of depigmentation have also been observed in Welsh Mountain ponies and, rarely, in Thoroughbred horses.

Aetiology/pathophysiology

- idiopathic or due to primary damage to melanocytes.
- acquired, possibly genetically programmed, depigmentation:
  - possible altered activity of the tyrosinase enzyme and production of anti-melanocyte antibodies.
  - autotoxicity has been proposed.
- compounds such as phenols may exacerbate pigment loss from affected melanocytes.

Clinical presentation

- depigmented circular spots up to 1 cm in diameter:
  - increase in number rather than size (Fig. 4.78).

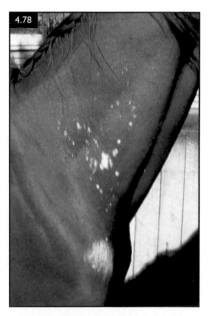

FIG. 4.78 Vitiligo depigmentation on the chest and neck of a 5-year-old Thoroughbred gelding. The number of areas had increased, but not their size, since the horse was a yearling.

- occasional white patches and leucoderma depigmentation.
- wax and wane in intensity but usually permanent.
- alopecia occasionally around the eyes.

Differential diagnosis

- copper deficiency
- equine CLE.
- leucoderma
- Appaloosa parentage.

Diagnosis

- clinical appearance and absence of injury are suggestive.
- history may indicate Arabian or Welsh Mountain pony heritage.
- leucoderma may not be synchronised with the area of associated leucotrichia.
- skin biopsy confirms the diagnosis:
  - early stages, mild lymphocytic infiltration with multifocal lymphocytic exocytosis.
  - later stages, complete absence of melanocytes.

Management

- treatment in humans includes topical/systemic calcineurin inhibitors (e.g. tacrolimus 0.1%), vitamin D, antioxidants and narrow-band UVB light.
- no reliably effective treatment in horses.
- topically applied or systemically delivered L-phenylalanine may help:
  - stimulate more pigment production.

Prognosis

- partial spontaneous recovery has been reported but is uncommon.
- number of spots may increase over time.
- condition is likely to be heritable – affected horses should not be used for breeding.

Hyperaesthetic leucotrichia

Definition/overview

- rare condition characterised by single to multiple crusts along the back.
- Arabians and their crosses, and American Paint horses, are predisposed.
- typically diagnosed in the summer months.
- Herpesvirus has been suspected as an underlying trigger factor.

- examination of lesions evokes a pain response:
  - pain can sometimes recur in subsequent years.
- lesions remain for 1–3 months and then regress:
  - followed by permanent white markings.
- occasionally outbreaks have been reported, suggesting an infectious trigger.
- no treatment.

Leucoderma (acquired vitiligo)

Definition/overview

- loss of pigment in the hair and skin can be related to various factors such as pressure sores, cryosurgery and surgery.

Aetiology/pathophysiology

- complication of many different diseases that destroy melanocytes, or that inhibit or change melanogenesis.

Clinical presentation

- white patches of hair and skin in irregular shapes may develop following:
  - surgery
  - X-ray radiation.
  - cryosurgery.
  - contact with irritants such as harness or rubber bits.
  - infection (Fig. 4.79).

FIG. 4.79 Leucoderma (acquired vitiligo) in a 12-year-old mare, which followed severe streptococcal infection of the mouth and eye region.

Differential diagnosis

- Equine coital exanthema.
- Arabian fading syndrome • equine SLE.
- Leucotrichia • Appaloosa parentage.

Diagnosis

- biopsies from affected areas reveals lack of melanocytes.

Management

- none known.

Prognosis

- pigment change is permanent.

Leucotrichia (tiger stripe, variegated leucotrichia, reticulated leucotrichia)

Definition/overview

- characterised by dorsal, bilateral, reticulated, white hair striping (Fig. 4.80).
- often seen in older horses with unknown earlier history:
  ○ occasionally yearlings are affected.
- Reticulated leucotrichia is indistinguishable from hyperaesthetic leucotrichia on biopsy, but clinically the lesions are not painful.

Aetiology/pathophysiology

- breed predisposition is apparent, indicating that genetic factors may be involved:
  ○ Standardbred and Quarter Horses are mainly affected.

Clinical presentation

- linear dorsal crusts are arranged in a cross-hatched pattern.
- temporary alopecia is present following shedding of crusts.
- new hair is white, but the skin remains with its original pigmentation.

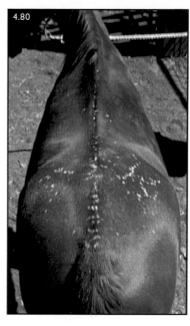

FIG. 4.80 Leucotrichia on the back line of a 6-year-old Standardbred mare. The white markings appeared at 2 years of age.

Differential diagnosis

- Leucoderma • Vitiligo.
- insect-bite reaction or hypersensitivity, or a drug- or vaccine-related trigger mechanism should be evaluated.

Diagnosis

- clinical history and examination help eliminate any other causes such as equine SLE.

Management

- no known specific treatments to stimulate pigmentation.
- eliminate trigger factors and use vitamin E supplementation to prevent further development.

Prognosis

- pigment change is permanent.

TRAUMATIC DISORDERS

Sunburn (actinic dermatosis) (see earlier)

Skin injury/Wounding
(see Chapter 5)

Urine/Faeces/Exudate scalding (see Book 3)

Physical and chemical skin injury

Definition/overview
- physical or chemical damage to superficial layers of skin leads to dermatitis and pruritus (Fig. 4.81).

Aetiology/pathophysiology
- application of chemicals to skin:
  - normal strength to horses with sensitive skin.
  - overstrength or irritant substances to skin.
- physical damage from burns, freezing, trauma and pressure necrosis from harness etc.:
  - mild to severe injury to the superficial to deep layers of the epidermis.
  - causing necrosis and skin death.

Clinical presentation
- varying degrees of itching ranging from:
  - mild and intermittent to severe with serious skin damage.
- fine to coarse scale and patchy alopecia.
- moist exudative dermatitis.
- may be painful to touch, particularly if secondarily infected.

Diagnosis
- visible dermatitis.
- history of injury, application of irritants or accidental application of inappropriate substances.

Management
- mild soap or shampoo wash with warm water, followed by clean warm water rinsing.
- gentle astringent can be used in mild cases.
- more severe cases:
  - silver sulfadiazine and/or corticosteroid topical ointment.
  - oral prednisolone, intravenous dexamethasone or pain-relieving medication to alleviate inflammation and discomfort.

Prognosis
- fair to good unless there is damage to the dermis, when permanent scarring and hair loss may result.

FIG. 4.81 Chemical-induced dermatitis following the application of overstrength insecticides. There was severe pruritus, alopecia with crusts and folding of skin over the withers. (Reprinted from Pascoe RR and Knottenbelt DC (1999) *Manual of Equine Dermatology*, WB Saunders, with permission.)

NEOPLASTIC CONDITIONS

Equine sarcoid

Definition/overview

- worldwide in all equid species including the horse, donkey, mule, ass and zebra.
- most common equine (cutaneous) tumour:
  - up to 90% of all skin tumours.
  - all areas of the body but some regions are less affected e.g. distal limb.
- locally aggressive tumour of fibroblast cells:
  - linked to Bovine papillomavirus (BPV) 1, 2 and 13.
  - multiple phenotypes of tumour and prominent secondary local effects.
- **no metastasis is recorded:**
  - local malignancy is relatively common in some anatomic regions:
    - side of the face, elbow and flank fold regions.
- distribution and phenotypes may vary from one geographical location to another:
  - may reflect feeding patterns of different species of flies in different seasons and geographical regions.
- 6 types are described but these are simply part of a continuum of clinical presentation (Fig. 4.82):
  - variable phenotype makes diagnosis more problematic as each type has its own set of differential diagnoses.
- not life-threatening unless neglected badly or have secondary implications:
  - significant limitations on use of the horse and loss of commercial value.
- very high rate of recurrence following:
  - unsuccessful treatment attempts (or biopsy).
  - inappropriate or irrational treatments.
- euthanasia is not uncommon:
  - where treatment cannot be assured.
  - cases are neglected and compromise the welfare of the horse.

Aetiology/pathophysiology

- BPV types 1 and 2 (and more recently BPV13 in Brazil) of subgroup A are usually associated with the disease.

- mechanism for disease and reason for variety of clinical presentations are uncertain.

Clinical presentation

- most sarcoids occur in sites where the skin is thin and hairless:
  - other sites usually involve some form of previous skin injury:
    - either fresh or partially or even fully healed.
- introduction of a horse with sarcoids can result in sarcoids in other previously uninfected horses on the farm within 6–8 months.
- may multiply or remain static for long periods.
- spontaneous regression and cure have been recorded.
- **only predictable thing about the equine sarcoid is its unpredictability.**

Occult sarcoid

Definition/overview

- hairless, often circular areas usually containing one or more small hyperkeratotic cutaneous lesions or micronodules.

Clinical presentation

- slow-growing, flat, circular, slightly thickened, scaly, hyperpigmented areas of alopecia (Fig. 4.83).
- may progress to small, warty verrucous growths within the surrounding skin:
  - usually become thickened or hyperkeratotic with time.
  - if injured, may develop into fibroblastic lesions.
- common on neck, face, sheath, medial thigh and axilla regions.

Diagnosis

- clinical features highly suggestive.
- can be mistaken for dermatophytosis, alopecia areata and pemphigus foliaceus.
- **collection of a biopsy sample may aggravate the lesion, inducing it to convert into an active fibroblastic sarcoid.**

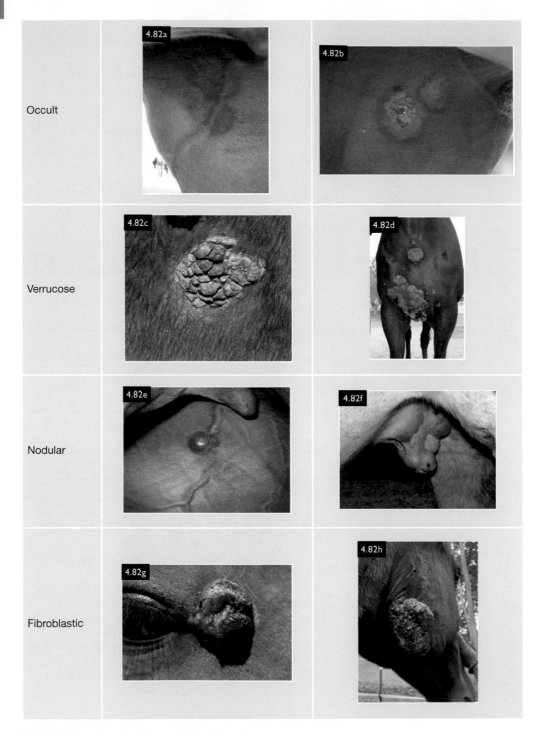

FIG. 4.82 Six categories of sarcoid *(Continued)*.

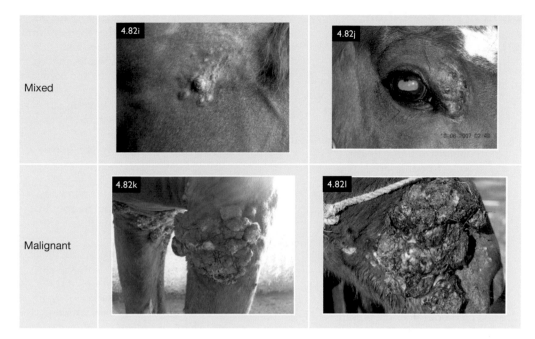

FIG. 4.82 *(Continued)* Six categories of sarcoid.

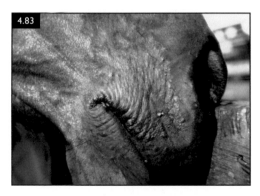

FIG. 4.83 Occult forms of sarcoid appear clinically as an area of alopecia containing one or more small nodules.

Verrucous (warty) sarcoid

Definition/overview

- usually slow-growing hyperkeratotic, wart-like lesions over variable areas:
  - small to extensive mainly around the face, neck, medial thigh and axilla.
- usually multiple and coalescent areas:
  - can be very extensive or in a miliary form with many small lesions.

Clinical presentation

- slow-growing, wart-like growths on the skin.
- variable sized lesions from tiny to extensive.
- sessile or pedunculated and quite frequently ulcerated in parts.

Diagnosis

- clinical features are usually pathognomonic:
  - often show a feint halo of occult change around them.
- easily mistaken for pemphigus foliaceous, papilloma and hypertrophic scarring.

Nodular sarcoid

Definition/overview

- cutaneous or subcutaneous, well-defined, firm nodules.
- single nodules, small numbers or extensive intertwining/coalescent groups of many nodules of variable sizes:
- erosion of overlying skin may be associated with more aggressive fibroblastic tumours.

Clinical presentation

- two separate types are found with profound differences in available treatment options:
  - ○ **Type A** does not involve the skin:
    - ♦ no attachment to overlying skin – moved independently over nodule(s).
    - ♦ may or may not be bound down to the surrounding tissues:
      - – **A1 nodules** freely moveable over the underlying tissues:
        - ◊ usually a loose fibrocellular capsule restraining tumour extension.
      - – **A2 nodules** have a bound-down nature:
        - ◊ cannot be moved independently of underlying tissues.
  - ○ **Type B** does involve the skin.
- subcutaneous (type A) and cutaneous (Type B) (usually) spherical nodules (0.5–20 cm) are evident.
- overlying skin may atrophy and split, revealing the nodule, which may then slough spontaneously.
- most frequent in the groin or eyelid area (Fig. 4.84).

Diagnosis

- biopsy is diagnostic:
  - ○ excisional biopsy preferred if a biopsy is considered necessary.

Fibroblastic sarcoid

Definition/overview

- fleshy often bleeding and exudative, exophytic appearance:

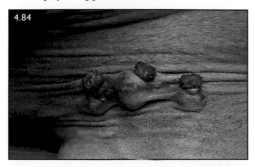

FIG. 4.84 Subcutaneous sarcoid nodules are often seen in and around the prepuce and scrotum.

- ○ very similar to granulation tissue but usually multifocal rather than uniform.
- more aggressive tumour stage and particularly dangerous when located on the face, throat, lower limbs and coronary band regions (Fig. 4.85).
- may follow minor/major wounds, surgical sites or injuries to other sarcoid types.
- divided into 2 subgroups:
  - ○ **Type 1 pedunculated** where there is a distinct pedicle:
    - ♦ **Type 1A:**
      - – pedicle does not contain any sarcoid tissue and no root in underlying tissue.
    - ♦ **Type 1B:**
      - – pedicle contains sarcoid tissue with a palpable/histologically detectable extension in the surrounding tissues.
  - ○ **Type 2** have a broad base with **no pedicle:**
    - ♦ usually, well-defined.
    - ♦ extensive broad-based extensions into the underlying and adjacent tissues.

Clinical presentation

- fleshy, ulcerative and exudative tumour resembling hypergranulation with erosion.
- most are locally invasive.
- pedunculated types (1A and 1B):
  - ○ recognised relatively easily.
  - ○ often difficult to identify clinically whether extensions exist beyond the lesion:
    - ♦ extensions may/may not be palpable in the pedicle and surrounding tissues.

FIG. 4.85 Fibroblastic sarcoid limb lesions are difficult to control and carry a less favourable prognosis.

- Type 2 (sessile) type are more broadly based and have no pedicle:
  - often develop rapidly after minor trauma of the other sarcoid types.
- surfaces are often ulcerated and exudative and these lesions are often complicated by:
  - superficial infection by bacteria (*Staphylococcus* spp.) or fungi (*Pythium* spp.).
  - invasion by larval myiasis and *Habronema* spp. larvae.

Diagnosis

- biopsy of non-ulcerated lesions is used for diagnosis.
- swab samples from the surface of ulcerated lesions subjected to PCR for BPV proteins is a poor and non-specific test.

Mixed occult, verrucous and fibroblastic sarcoid

Definition/overview

- progressively more aggressive, as changes from occult and verrucous types to the fibroblastic type.

Clinical presentation

- probably a progressive state from either an occult or a verrucous type.
- all three entities may be evident (Fig. 4.86).

Diagnosis

- presence of more than one form of sarcoid is almost pathognomonic.

Malignant/Malevolent sarcoid

Definition/overview

- recently identified variation showing increased local malignancy.
- no metastasis but local invasion can be severe and extensive (Fig. 4.87).
- subtype involves the malignant nodular expanding ring form of the disease.

Clinical presentation

- may result from other types of sarcoid subjected to repetitive injury, medication or incomplete surgery.
- occasionally, with no previous history of interference, they spontaneously develop into multiple tumours locally and/or metastasise to lymphatics and lymph nodes:
  - cords of tumour extend into the lymphatics.
- particularly dangerous in the periocular region.

Diagnosis

- changes to a more benign sarcoid type are a common feature.
- invasive regrowth from a previously incorrectly treated sarcoid.

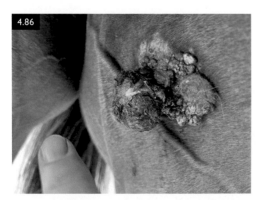

FIG. 4.86 This mixed sarcoid has elements of occult, verrucose, nodular and fibroblastic sarcoid with no dominant type. Note the relationship to the horizontal stifle vein which adds complexity and unpredictability.

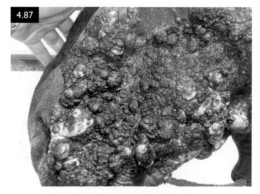

FIG. 4.87 Example of malignant sarcoid on the lower face.

Sarcoid: general management considerations

- owner should be made fully aware of the serious nature of this disease.
- early intervention is better than delayed intervention.
- full discussion of evidence-based treatment options:
  - likelihood of successful treatment and prolonged or repeated treatments.
  - palliative management discussed where ideal treatments are not available.
- benign neglect only considered if case beyond treatment or no available treatments are likely to resolve the disease:
  - partial management (interference) often results in exacerbation.
  - spontaneous full and permanent self-cure of single lesions does occur.

Treatment options

- before embarking on any treatment, the following should be considered:
  - correct diagnosis:
    - confirmation by biopsy or excisional biopsy may be required.
    - classification of the type for each lesion is important.
  - how large is the tumour and its extent and depth.
  - where is the lesion anatomically:
    - some sites are more/less tolerant of different treatment options.
  - how long has the lesion been present:
    - recent or chronic
    - fast-growing or slow expansion.
  - previous interference or treatment and when:
    - previous unsuccessful treatment reduces the prognosis for all subsequent treatments on all lesions in the horse.
    - best available treatment should be used first.
  - likely prognosis for the lesion and the horse:
    - poor prognosis can be managed in a palliative manner.
  - logistics of the treatment:
    - is it practical.
    - single or multiple treatments.
    - length of time under treatment.
    - horse stabled or hospitalised.
    - facilities and expertise available.

- patient compliance – tolerate treatment or associated management.
  - cost of the treatment in respect of the 'value' of the horse.
  - owner compliance:
    - owner fully understands treatment method and its implications.
  - health and safety issues associated with the treatment method.

> A compromise is almost always required but, in general, the best possible treatment should always be used on the first occasion.

Benign neglect

- means the lesion(s) will be monitored at suitable intervals, often with the cooperation of the owner to manage the case for a defined period:
  - any deterioration of the individual lesions will trigger immediate treatment.
- **always better to treat early neoplastic lesions than later ones.**

Spontaneous resolution

- cases can completely resolve spontaneously without any treatment or other interference sometimes over a limited time scale (usually weeks or months):
  - further lesions do not seem to occur implying some immunological response that protects the case in spite of its genetic susceptibility.
  - all lesions may resolve or only individual lesions whilst others that are present develop or remain static.
- impossible to predict which lesions will undergo spontaneous resolution.
- unwise to rely on this as a method of management.

Ligation

- elastrator (lamb castration) rings, Lycra or even heavy elastic bands can be used but are potentially problematic.
- can only be used with type A1 nodule or where other concurrent therapy is used:

- single pedunculated sarcoids with loose skin on the body or neck.
- proper placement of ligature (i.e. skin-only pedicles).
- lesion extends beyond the ligature, then recurrence will occur.
- easy to apply.

Surgical removal

- removal of the sarcoid, plus at least 15–20 mm of normal tissue around the lesion:
  - using standard sharp surgery or laser (diode/CO_2) or harmonic scalpel.
  - under local or general anaesthesia.
  - can be effective.
  - surgical pathology is compulsory to establish the safety of the excisional margin.
- advantages include:
  - full control of tissue removed and wound closure/grafting of site.
  - fast, one-off procedure.
  - opportunity for concurrent procedures including immunotherapy, radiotherapy and chemotherapy.
  - rapid healing if the surgical margins are safe:
    - residual sarcoid tissue or seeded tumour cells released into the wound during the procedure is a common source of failed surgical outcomes.
    - mitigated by careful surgical procedures and haemostasis.
- disadvantages are:
  - recurrence rate of over 50% (up to over 90%) dependent on case selection.
  - failure reduces the prognosis:
    - surgical site may act as a source of sarcoid for other sites and even other horses.
  - removal of large areas of normal skin precludes closure and slows healing time.
  - laser sites heal slower than other surgical procedures.
  - inability to remove this quantity of skin from limb lesions increases the risk of recurrence.

Electrocautery

- often successful when used on defined small, superficial lesions in convenient

locations where margin definition is clear and where an extra margin can be removed:
  - recurrence is common.
  - all excised tissue must be submitted for surgical pathology.

Cryosurgery

- useful but depends on correct technique:
  - significantly enhanced by using concurrent chemotherapy either topically or by intralesional injection.
  - useful when applied after surgical excision (partial or complete).
  - three freeze–thaw cycles should be used.
  - complications include:
    - under- or overtreatment.
    - injury to surrounding blood vessels, nerves, bone and tendon.
  - leucoderma and leucotrichia will occur at the treated site.

Radiation therapy

- gold standard for treatment carrying the highest success rate of all treatments.
- Teletherapy/external beam radiation requires expensive and highly controlled environments so is rarely accessible for horses:
  - very effective.
  - invariably requires GA.
- Brachytherapy involves implanting radioactive materials inside the tumour:
  - usually iridium[192] wires or pellets/seeds:
    - gold-standard treatment for sarcoid and also to varying degree in carcinoma, soft-tissue sarcoma, cutaneous lymphoma and melanoma.
- Plesiotherapy involves the placement of a radioactive source (usually strontium[90]) directly over the tumour site for several days:
  - only suitable for superficial small lesions such as along the eyelid margin.
- all techniques can be combined with surgical 'debulking' or chemotherapy before or after the radiation.
- often used as method of last resort when other methods have failed.
- advantages of radiotherapy include:

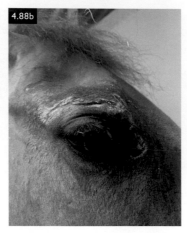

FIG. 4.88 (a) This pony gelding developed an aggressive invasive fibroblastic sarcoid in its right upper eyelid. Combination of intralesional chemotherapy using carboplatin emulsion and high-dose-rate (HDR) radiation resulted in a very pleasing outcome without any apparent functional limitations **(b)** at 7 months post-treatment.

- ○ high efficacy and very high accuracy with less functional compromise on adjacent normal tissues (Fig. 4.88).
- ○ most treatments are now performed under standing sedation.
- ○ reduced scarring and disfigurement.
- • disadvantages of radiotherapy include:
  - ○ full effects not delivered before 12 months after completion of the course.
  - ○ specialist facilities and staffing required.
  - ○ expensive.

Photodynamic therapy (PDT)

- • PDT applications involve:
  - ○ photosensitising agent such as Aminolaevulinic Acid or Hyperacin administered in the absence of light, either systemically or topically.
  - ○ photosensitiser is activated by exposure to light for a specified period from a light source capable of delivering a specific wavelength of light.
  - ○ reactive oxygen kills the target cells.
- • objective is to excite the photosensitiser to produce radicals and/or reactive oxygen species within the cells (phototoxicity).
- • scientific and anecdotal reports of efficacy against sarcoid.
- • equipment is specialised and requires experience to use effectively.

Radiofrequency hyperthermia

- • reported to be effective in some sarcoid and carcinoma cases.
- • most successful with small tumours but repeat treatments are commonly required.
- • not gained wide acceptance.

Chemotherapy

- • varied cytotoxic and antimitotic chemicals have been used to treat sarcoids.
- • can be delivered topically or by intralesional injection:
  - ○ systemic chemotherapy used for very extensive lesions but logistically challenging and has special requirements.
- • **topical medications include:**
  - ○ 5% 5-fluorouracil cream applied according to schedule depending on the type, location and severity of the lesion itself.
  - ○ **AW5 cream:**
    - ♦ mixture of heavy metal salts and antimitotic compounds.
    - ♦ aggressive, cheap and effective method of treatment for a wide range of sarcoid types in different anatomic locations.
    - ♦ restricted availability.
  - ○ **Imiquimod 5% topical cream (Aldara, 3M):**
    - ♦ antiviral antitumour compound with limited beneficial effects.
    - ♦ most applicable to small lesions on the face or ear.
    - ♦ very painful but useful method in some circumstances.
    - ♦ expensive for large lesions.
  - ○ **mixture of Zinc Chloride (usually 12–25%) with extracts of blood root (*Sanguinaria canadensis*):**
    - ♦ reports of successful outcomes using various commercial forms.
    - ♦ used with caution and only under veterinary supervision.

Electrochemotherapy

- based on the increasing permeability of cell membranes to some chemotherapeutic compounds when the cells are subjected to controlled electrical currents (electroporation):
  - aqueous chemotherapeutic such as cisplatin is injected into the lesion immediately prior to the application of the electrode probe and the delivery of the electrical pulses.
  - several repeat treatments (up to 6 or more) and each one requires deep GA.
- little or no advantage over alternatives that do not require repeated GA.

Immunotherapy (vaccination or modulation of the immune system)

- **Intralesional Bacillus Calmette–Guérin (BCG) therapy:**
  - evidence-based efficacy particularly for eyelid lesions.
  - less effective elsewhere and contraindicated on the limbs.
  - repeat use of BCG and its analogues carries a significant risk of anaphylaxis.
- purified protein derivative of **mycobacterial cell wall fractions** (Immunocidin®, NovaVive, USA):
  - emulsion modified to reduce the toxic and allergic effects whilst retaining antitumor activity.
  - like BCG, activates macrophages and T-cells and induces tumour destruction.
  - suggestion of a remote effect on other tumours in the horse.
  - response can be very strong.
  - dose dependent so bigger tumours get bigger doses.
  - expensive but can be very effective in some selected cases.
- **Tigilanol tiglate** (Stelfonta, Virbac, USA):
  - used to treat isolated individual tumours.
  - maximum permitted dose is easily reached and only used under very strict control.
  - pain reactions can be immediate and extreme but settle quickly.
  - site sloughs widely and heals remarkably quickly.

- Autologous vaccines are irrational and should not be used.

Prognosis

- varies from good to poor depending on the site, number and type of lesions, and response to initial treatment.
- some 'Rules' of the sarcoid that should be considered in all cases:
  - a horse that has even one sarcoid is liable to the disease:
    - ♦ likely to be genetic in origin and remains with the horse for life.
    - ♦ new lesions and recurrences at previously treated sites are to be expected.
  - sarcoids appear to multiply over the summer months (fly seasons) and grow over the winter:
    - ♦ likely to be vector related.
    - ♦ fly control should be instituted.
    - ♦ affected horses quarantined behind netted stable doors and windows during the times of the day when flies are active.
  - the more lesions a horse has, the more they get, and equally the opposite:
    - ♦ all sarcoids should be treated as early, thoroughly and effectively as possible.
    - ♦ treatment prior to spring turnout is logical.
  - all treatments are problematic and none is 100% effective:
    - ♦ treatments need to be adjusted to the specific needs of the case/lesion.

Neurofibroma (Schwannoma)

Definition/overview

- benign nerve-sheath tumour associated with the peripheral nervous system.
- tumour of Schwann cells.
- most equine Peripheral Nerve Sheath Tumours (PNST) are solitary and histologically almost indistinguishable from sarcoid:
  - many cases are ultimately considered to be sarcoid or sarcoid-like tumours with similar behavioural characteristics.

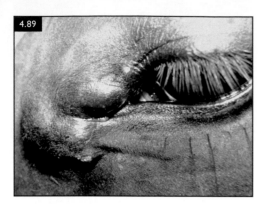

FIG. 4.89 Neurofibroma is most frequently found in the upper and lower eyelids as small hard subcutaneous nodules, with hairless areas around the tumour, similar to sarcoid.

- most often found in the upper and lower eyelids (Fig. 4.89) and in the sheath region.

Aetiology/pathophysiology

- cause is unknown.
- suggestive of a common aetiology with sarcoids in some cases.

Clinical presentation

- progressively enlarging solid, usually spherical nodular lesion(s) with slow expansion:
  - often with late development of some surrounding alopecia and scaling.
- solitary or multiple lesions can be found.
- can reach a considerable size and can be irregular.
- some nodules are inclined to ulceration:
  - dramatic proliferation growth of nodules causes erosion through overlying skin.
  - development into a broadly based fibroblastic-type lesion.
  - seldom pedunculated.

Diagnosis

- probably impossible to clinically differentiate from a sarcoid.
- ultrasonography can be helpful:
  - usually very homogeneous.
  - sometimes have a granular echogenicity and show marked heterogeneity.
- excisional biopsy of non-ulcerated areas and histopathology are usually diagnostic:

  - diagnosis has to involve the absence of BPV genomic markers characteristic of sarcoid.
  - sarcoid, neurofibroma, low-grade fibrosarcoma and fibroma are often difficult to differentiate unless there are clinical features that assist the diagnosis.
  - immunohistochemistry is often used to clarify the diagnosis.

Management

- complete surgical removal is required as soon as possible:
  - local recurrence occurs in 25–50% of all operated cases.
  - sometimes some distance from the 'parent' lesion.
- intralesional injection of BCG has been successful.
- local slow-release chemotherapy methods may be successful.
- radiotherapy has been quite effective in some cases.

Prognosis

- guarded.

Fibroma/Fibrosarcoma

Definition/overview

- very rare tumour of fibroblast cells.
- often single, firm or soft, well-circumscribed, dermal or subcutaneous nodules but some are very wide and invasive.
- may ulcerate or develop into a flattened verrucous (warty) lesion.
- Fibromas are benign.
- Fibrosarcomas may be multiple, locally invasive and malignancy can occur.

Aetiology/pathophysiology

- arise from dermal or subcutaneous fibroblasts.

Clinical presentation

- dermal or subcutaneous nodules, which may ulcerate (Fig. 4.90) or develop into verrucous to cauliflower-like tumours on the head, limbs, neck and flanks.
- highly invasive and possibly malignant forms do occur.

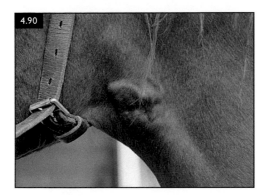

FIG. 4.90 Fibroma. A slow-growing, flat fibroblastic tumour on the lateral neck. Histopathology is required to confirm the diagnosis.

FIG. 4.91 Typical solar fibroma in the frog region of a 16-year-old gelding causing some discomfort but no overt lameness. Histological examination confirmed fibroma. Repeated surgical interventions were required to control it.

- solar fibroma is the most common neoplasm of the frog tissue of the foot (Fig. 4.91).

Diagnosis
- histopathology of biopsy sample or an excisional biopsy is essential to confirm margins.

Management
- total surgical excision is required but is very difficult to achieve:
  - any residual tumour tissue will simply recur.
- cryotherapy is usually only partly successful even when combined with local chemotherapy.
- radiation is more successful but seldom available.

Prognosis
- all forms of fibroma or fibrosarcoma guarded at best.
- recurrences requiring repeated interventions of various types are very common.
- malignancy carries a poor/hopeless prognosis.

Melanoma/Melanocytoma

Definition/overview
- tumour of melanocytes occurring more commonly in grey- and white-coated horses:

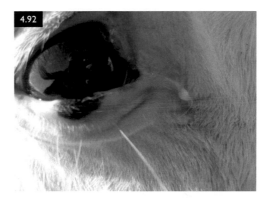

FIG. 4.92 This 12-year-old pony mare was presented for investigation of an ocular discharge. The primary problem was an upper eyelid carcinoma but, incidentally, an melanotic naevus was noted in the lower lid. The latter had been unchanged since it was first noted at purchase some 9 years previously.

  - other colours can also be affected.
- several classification schemes have been proposed to correlate histopathological appearance with clinical behaviour (i.e. benign or malignant).
- 3 basic types of melanocytic skin tumours have been suggested:
  (1) melanocytic nevi (melanocytoma) (Fig. 4.92).

FIG. 4.93 Typical example of dermal melanoma; these are usually solitary and well defined but can be coalescent and merge into larger areas of affected skin. As in this case, many lesions are seen in early-stage development.

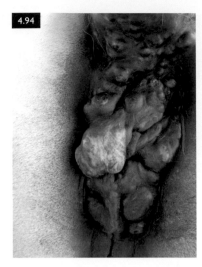

FIG. 4.94 A typical aggressive malignant melanoma. Notice the loss of colour indicating non-differentiated cells. Metastatic tumours were found in the mediastinal lymph nodes, the heart valves and the spleen. Histological malignancy was identified in several of the perineal sites.

(2) dermal melanoma (Fig. 4.93).
(3) anaplastic malignant melanoma (Fig. 4.94).

Aetiology/pathophysiology

- both benign and malignant tumours arise from melanocytes:
  - genetic factors are likely to be heavily involved in forms occurring in grey horses.
  - breeds such as the Lusitano, Andalusian and the Lipizzaner are often severely affected.
  - Arabian and Thoroughbred horses and other breeds are also commonly affected.
  - grey donkeys and Eriskay ponies (most are grey) are extremely rarely affected.
  - no suggestion they are related to sunlight exposure:
    - most occur in sites where sunlight exposure would never occur.
  - malignancy is common with 70% or more of tumours developing to some degree if left long enough:
    - fortunately, the process of dissemination is extremely inefficient.

- malignant cells in the blood stream are filtered out by the pulmonary capillary bed where they do not expand or grow.
- cells that escape the pulmonary filter cause metastatic spread with effects in:
  - central nervous system.
  - pleural and peritoneal cavities.
  - parotid salivary glands.
  - muscles.
  - all the major internal organs.
  - chronicity is an important feature of advanced forms and larger tumour complexes.
  - malignancy is not related to size of tumour (Fig. 4.95).

Clinical presentation

- some breeds and families of grey horses have a higher incidence of melanoma.
- initially small hard cutaneous tumours that slowly increase in size.
- primary tumours are commonly found around the anus, vulva, tail and prepuce, the lips and eyelids, and on the skin in general (Fig. 4.96).
- lesions in the parotid salivary glands are secondary/metastatic tumours (Fig. 4.97):

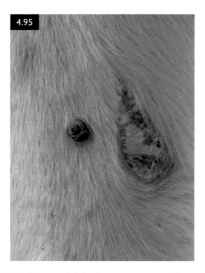

FIG. 4.96 Melanocytoma with benign-looking nodules around the anus and vulva exuding melanin. Ulceration should be taken to indicate a significant change in the cell behaviour.

FIG. 4.95 4-year-old Arab cross mare with a small black tumour on her right forelimb. The tumour was routinely removed and found to have a very high malignancy. No further lesions developed, and no consequences arose for up to 5 years when lost to follow-up.

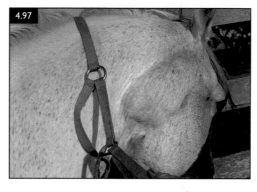

FIG. 4.97 Melanoma in a 12-year-old Welsh Mountain pony with a secondary malignant tumour in the right parotid gland area that was inoperable.

- o indicate careful examination of the guttural pouches.
  - o possibility of disseminated tumours in other sites with functional compromise.
- single masses are rare and usually multiple tumours of varying size and pathological behaviour are present in most cases, particularly in grey horses:
  - o single tumour may occur in 'other'-coloured horses.
  - o usually with a more aggressive appearance and difficult prognosis (Fig. 4.98).
- malignant melanoma is rapidly invasive and responds poorly to surgery or chemotherapy:
  - o may ulcerate and ooze black exudate mixed with variable volumes of blood.
- Melanocytomas are 'depots' of melanin pigment, non-invasive and cause space-occupying disruption such as perineal/pelvic obstruction through expansion.

Diagnosis

- diagnosis of melanoma is a 'broad brush' as the single name includes a wide variety of pigmented tumours, some of which are benign and others malignant.

- most common tumour found in grey horses, but it is not the only tumour:
  - o mixtures of melanoma and sarcoid or even carcinoma also occur.
- visible tumours should be examined carefully.
- guttural pouch endoscopy should be performed on all horses with significant melanoma lesions as this is the only site where dissemination can be seen directly (Fig. 4.99):
  - o particular attention paid to lateral wall and roof of the lateral compartment.
- biopsy and histopathology are diagnostic although many pathologists are uncertain of the degrees of malignancy:

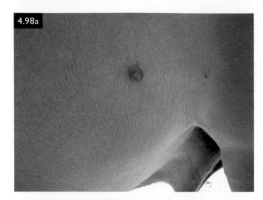

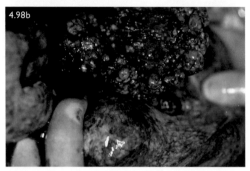

FIG. 4.98 (a) This bay horse developed a dark tumour as shown over some months. Histological examination showed it to be a very high malignancy melanoma. It was successfully removed, and no other tumours were seen for 3 years. **(b)** The horse was then presented for colic and its whole abdomen was filled with myriad melanoma lesions some of which were black and others of a cream or blue-grey colour.

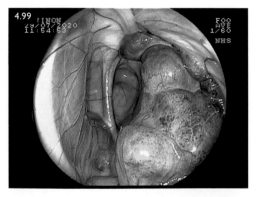

FIG. 4.99 A dramatic melanoma in the wall of the guttural pouch of a middle-aged gelding with a long history of multiple melanoma lesions in the lips, the perineum and sheath regions.

o differentiation between melanocytoma and melanoma is not always clear histologically and probably should be treated as variations of same condition.
o melanocytic naevus is the only variant that can safely be ignored as they appear not to develop into melanoma, expand or show malignancy.

Management

• varies with the type and location of the tumour and the degree of malignancy:
  o latter is poorly understood but most tumours start off as benign, accessible skin masses and culminate in some degree of malignancy if left long enough.
  o treatment at an early stage is therefore preferred.
• surgical or laser surgical or diathermy excision of:
  o lip, preputial, perineal, perianal, perirectal or ventral tail lesions:
    ♦ often carries a good outcome even if small areas of residual tumour are left.
    ♦ sites heal well with minimal clinical interference.
    ♦ post-surgical chemotherapy may help to minimise recurrence:
      – slow-release forms of cisplatin (usually in biodegradable beads).
      – carboplatin in a slow-release emulsion.
• cryosurgery (3 freeze–thaw cycles) +/– intralesional, slow-release chemotherapy such as carboplatin or 5-fluorouracil can be used in conjunction with surgical excision or in various combinations.
• vaccination with a DNA plasmid vaccine encoding human tyrosinase (ONCEPT®, Boehringer Ingelheim Animal Health, USA):
  o appears to be safe and well tolerated.
  o requires intradermal injection using a 'needleless' transdermal device (usually) in the pectoral region.
  o every 14 days for a primary course of 4 procedures followed by a 6-month booster and then 6-/12-month boosters indefinitely.
  o expensive and should only be regarded as delaying progression, not as a cure.

Prognosis

- no cures for this disorder and the prognosis for all cases is guarded.
- recognisable indicators of early malignant development are rarely present and so most cases with disseminated melanoma are recognised late.
- prognosis probably significantly improved by early removal of all (or as many as possible) of the smaller masses in the skin:
  ○ repeated surgical removal carried out every year or two.
  ○ good overall effect and improves the short-, medium- and long-term prognoses.
- prognosis for non-grey horses is best regarded as very guarded:
  ○ usually noticed earlier and occur in limited numbers leading to earlier removal.

Squamous cell carcinoma

Definition/overview

- second most common tumour of horses comprising approximately 20% of equine neoplasms in most studies.
- breed and colour predilections are recognised.
- most commonly affected include draughts such as Shire and Clydesdale, Arabians, Appaloosas, and American Paint and Pinto horses.

Aetiology/pathophysiology

- malignant tumour arising from keratinocytes.
- exposure of non-pigmented skin to actinic (solar) radiation is frequently involved in facial, eyelid or conjunctival carcinoma.
- high UV radiation in tropical regions or higher altitudes may increase the incidence.
- papillomavirus-related carcinogens are often implicated in vulvar and penile/preputial carcinoma.
- older horses often have penile or vulvar warts, and these may be a precursor:
  ○ smegma accumulations within the prepuce is probably less significant than originally thought (Fig. 4.100).
- geldings are more often affected by penile carcinoma than entire/working stallions.

Clinical presentation

- small cutaneous ulcerated/granulating skin defect that may be depressed below skin level.
- may have a proliferative or destructive nature.
- most carcinoma sites have a characteristic malodour even with early lesions.
- commonly affects the non-pigmented mucocutaneous junctions:
  ○ ocular conjunctiva (see Book 5) and the external genitalia (see Book 2):
    ♦ early incipient carcinoma usually appears as slightly raised white plaques on the penile skin.
    ♦ vulvar, clitoral, penile and preputial carcinoma may vary from proliferative cauliflower-like to destructive/erosive lesions:
      – areas may bleed easily.
    ♦ eye and eyelid lesions often start as white, raised plaques at the edge of the lid or corneal/scleral junction:
      – may progress rapidly to granulomatous and ulcerated lesions.
      – corneal *in situ* carcinoma is occasionally encountered.
  ○ other areas of the body that are sometimes affected include the perianal skin, mouth, lips, nostrils and skin at the perioral and perinasal regions:
    ♦ vulva and anus masses are often slow-growing.
    ♦ around nose and mouth appear as a depressed ulcer (Figs. 4.101, 4.102):

FIG. 4.100 Squamous cell carcinoma of the prepuce. This depigmented skin tumour was moveable with preputial skin and was safely removed surgically.

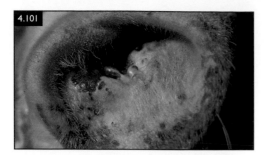

FIG. 4.101 Squamous cell carcinoma at the muco-cutaneous junction of the nose.

FIG. 4.102 Squamous cell carcinoma with meta-static spread to the jowl area from the nose of a 25-year-old Arabian stallion. The horse was euthanised.

- progresses to a granulomatous and malodorous growth/erosion.
- local metastasis to the lymph nodes has been noted in 18.6% of cases:
  - regional lymph nodes should be assessed by palpation, inguinal and rectal examination, and (ultrasound-guided, if available) fine-needle aspiration.
  - may also spread to the lungs.
- carcinoma can develop in wounds, causing problems with the healing process.

Diagnosis

- appearance of the carcinoma should provide at least a tentative diagnosis.
- diagnosis confirmed histologically by biopsy or excisional biopsy.

Management

- early diagnosis is key to successful treatment:
  - early metastatic spread does occur and can become apparent later.
- single-modality treatments seldom work and combined surgical debulking and post-surgical chemotherapy is recommended.
- surgical excision:
  - penile amputation or en bloc resections can be effective in advanced cases.
- CO_2 laser ablation or cryosurgery.
- post-op chemotherapy including:
  - 1% 5-fluorouracil ocular drops, creams or intralesional slow-release forms.
  - 5% 5-fluorouracil cream is also very useful:

- penile carcinoma – applied once weekly for 5–7 weeks:
  - usually in conjunction with cryosurgery or sharp surgical debulking.
- other skin sites – applied twice daily on an ongoing basis:
  - combine with cryosurgery, intralesional chemotherapy with carboplatin in slow-release form or cisplatin biodegradable beads.
  - slow-release carboplatin can also be injected around the surgical site and repeated at intervals depending on the pathology results.
- radiotherapy using strontium[90] plesiotherapy (ocular) or iridium[192] brachytherapy:
  - good to excellent long-term outlook.
  - considered the gold standard.
- recurrence is common, especially if predisposing factors cannot be eliminated:
  - solar radiation, smegma and trauma.
- thoracic metastasis presents with progressive weight loss, anorexia and intermittent fever and should be confirmed by chest radiographs:
  - significant paraneoplastic signs associated with this malignancy.
  - metastatic pulmonary tumours carry a hopeless prognosis.

Prognosis

- guarded prognosis always given.
- signs of metastatic spread may be very delayed (up to 5 years or more).
- early wide surgical excision can lead to complete remission (Fig. 4.103):
  - amputation of the tail or enucleation of the eye.
  - amputation of the penis or clitoridectomy coupled with chemotherapy may be successful.
  - surgical removal of long-standing invasive tumours on the eyelids, prepuce, mouth and, to a lesser extent, the vulva is less successful.

Mast cell tumour/Equine cutaneous mastocytosis

Definition/overview

- debate about whether these are genuine mast cell tumours as they behave in a very different way to other species.
- usually, a single cutaneous nodule characterised by:
  - focal aggregation of mast cells and large numbers of eosinophils.

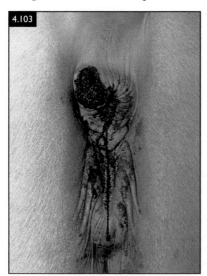

FIG. 4.103 A tumour involving the anal sphincter in an 8-year-old piebald pony, which was surgically removed. There was no return in 3 years. Histopathology was required to confirm diagnosis of a slow-growing squamous cell carcinoma. (Reprinted from Pascoe RR and Knottenbelt DC (1999) *Manual of Equine Dermatology*, WB Saunders, with permission.)

  - fibrinoid collagen necrosis and, occasionally, mineralisation of the contents is present, giving it a gritty texture.
- multiple lesions can occur, and these are less responsive to treatment.

Clinical presentation

- 1–18 years of age (mean, 9 years) without sex or breed predilection.
- single, firm, well-defined nodules 2–20 mm in diameter:
  - occurring most commonly on the head and limb regions.
  - common form is found in the nasal mucosa within the nostrils (Fig. 4.104).
- surface may be normal, hairless or ulcerated.
- some lesions are pruritic and induce various degrees of self-trauma.
- multiple mast cell tumours can occur in newborn foals:
  - these usually regress spontaneously.

Diagnosis

- fine needle aspiration or biopsy via hollow needle, punch or excisional biopsy:
  - examined by a histopathologist.
  - most mast cell tumour lesions have a very dominant eosinophil content.
- ultrasonography is useful as there is an almost pathognomonic heterogeneity to image created by the calcified areas within the tumour.

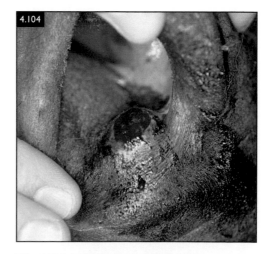

FIG. 4.104 Mastocytoma. There is an ulcerated lesion in the ventral nares of this horse.

Management

- surgical removal is usually curative even if margins are not clear:
  - recurrence is uncommon.
- intralesional corticosteroids beneficial for areas where surgery may be difficult or disfiguring, or useful to reduce the size of the tumour prior to surgical excision.

Prognosis

- generally good to guarded.

Haemangioma/ Haemangiosarcoma

Definition/overview

- uncommon tumours.
- found most often in the elbow/axillary region (Fig. 4.105), groin, thorax and distal limbs.

Aetiology/pathophysiology

- benign (angioma) or malignant (angiosarcoma) neoplasm arising from vascular endothelial cells.
- locally invasive and potentially metastatic.

Clinical presentation

- often present with numerous tortuous and enlarged blood vessels.
- may ulcerate and bleed very easily – often spontaneously.

Diagnosis

- biopsy is essential and histopathology is diagnostic.
- potentially a dangerous tumour so early diagnosis is essential.

Management

- complete and urgent surgical removal is essential:
  - best performed under GA as bleeding is often severe and the margins of the tumour are hard to define.

Prognosis

- benign forms – fair.
- progression and recurrence are usually associated with transformation to malignancy:
  - very poor prognosis.

(Malignant) histiocytoma (giant cell tumour of the soft parts)

Definition/overview

- rare tumour of histiocyte cells.

Clinical presentation

- solitary, poorly circumscribed masses of variable size.
- many are firm but if ulcerated become fragile and spongy in texture (Fig. 4.106).

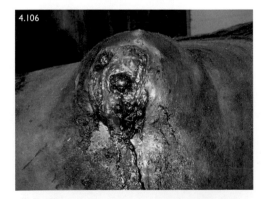

FIG. 4.106 15-year-old gelding developed a small, firm nodular mass on the dorsal thorax which expanded slowly over 6–9 months, It then ulcerated and grew fast. The tissue was spongy and easily disrupted resulting in profuse bleeding. Surgical removal led to a good outcome.

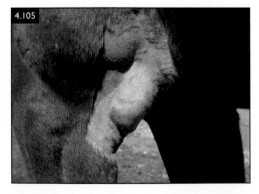

FIG. 4.105 Haemangioma in a 3-year-old Standardbred. A fluctuant swelling with enlarged blood vessels is present around the elbow area.

- most frequently occur on the neck and limbs.
- locally invasive and slow to metastasise.

Management
- complete surgical excision.
- does not respond well to local chemotherapy.

Prognosis
- generally poor and recurrence is common due to the locally invasive nature of the tumour.

Lymphoma

Definition/overview
- Equine cutaneous lymphoma is divided into two categories:
  - **cutaneous T-cell lymphoma** (CTCL):
    - epitheliotropic generalised scaling dermatosis.
    - carries a grave prognosis.
  - **T-cell-rich, large B-cell cutaneous lymphoma** (TCRLBCL):
    - non-epitheliotropic solitary to multiple nodules.
    - affects entire body with some predilection for the eyelids.
    - survival time post-diagnosis varies from months to years.
    - age of onset ranges from 2 months to 31 years (mean, 10.7 years).
    - some cases have concurrent endocrine tumours that appear to exacerbate the lymphoma e.g. granulosa (thecal) cell tumour (GCT):
      - removal of the GCT results in significant (seldom complete) resolution.

Clinical presentation
- most cases present with cutaneous or subcutaneous masses of varying size and in different locations (Fig. 4.107):
  - may have localised or generalised lymphadenopathy.
  - subcutaneous spread, slight discharge and some crust formation
  - scaling over the surface of the skin often part of the mycosis fungoides

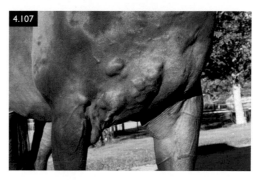

FIG. 4.107 Widespread cutaneous and subcutaneous nodules developed in this 8-year-old Warmblood gelding following mild weight loss and reduced performance. A diagnosis of T-cell-rich B-cell lymphoma was made after biopsy and immunohistochemistry.

syndrome associated with T-cell lymphoma.
- Paraneoplastic syndrome often the first sign and includes:
  - low-grade persistent and progressive but non-responsive anaemia.
  - hypercalcaemia and undulant fever of unknown origin:
    - often misinterpreted or overlooked.
- early presentation of cases gives a much better chance of a successful treatment.

Diagnosis
- biopsy and immunohistological confirmation that the nodules are composed of:
  - densely packed lymphoblastic cells expressing CD79a.
  - numerous small, round, CD3-positive T lymphocytes.
- abnormal circulating lymphocytes are seen in 25–50% of cases.
- overt leukaemia is very rare.

Management
- options largely depend on the nature of the lymphoma:
  - T-cell forms are very serious and require systemic chemotherapy:
    - specialist requirement.
  - B-cell lymphoma has a slower and more manageable course:

- ◆ use of combinations of prednisolone and cytarabine (also known as cytosine arabinoside), methotrexate, cyclophosphamide and other drugs that constrain lymphocyte metabolism.
  - ◆ prednisolone carries a reasonable clinical benefit on its own and is easier to manage than some of the other drug combinations.
- treatment regimens for the management of lymphoma in horses are published in specialist medicine texts.
- Progestins can help since many are, to some extent, exacerbated by oestrogens.
- surgical removal of ovarian tumours should be undertaken if present.

Prognosis

- generally, very poor compared to other species:
  - ○ highly dependent on the type of tumour and the stage of presentation:
    - ◆ late presentation: often because the condition is not recognised.
  - ○ paraneoplastic syndromes often the most serious and must be addressed.
  - ○ cases with regional/generalised lymph node enlargement, rapid weight loss, lethargy, ventral oedema and pyrexia are probably hopeless.
  - ○ other primary forms of lymphoma in other organs (thymus, mediastinum, spleen or liver) or metastatic tumours in other organs carry a poor prognosis.

Wound Management and Infections of Synovial Structures

Classification of wounds

- **open** (complete epithelial loss) or **closed**:
  - open wounds include lacerations, incisions, ulcers and punctures (Fig. 5.1).
  - closed wounds include contusions, abrasions and haematomas (Fig. 5.2).
- defined by cause, subsequent pathophysiology and anatomic location:
  - causes include mechanical, thermal, chemical, or subsequent to irradiation.
- important to assess:
  - vascular supply to the damaged area.
  - degree of contamination.
  - possibility of infection.
  - involvement of adjacent anatomic structures.
- lacerations and puncture wounds are particularly common:
  - following wire- or wood-fence injuries:
    - ♦ barbed wire can lead to:
      - severe loss of tissue (degloving injuries) (Fig. 5.3).

- exposure of large amounts of connective tissue and bone.
    - ♦ smooth-wire lacerations can lead to:
      - vascular strangulation of the affected extremity.
- closed wounds:
  - may result from kicks or self-inflicted trauma.
  - abrasions present with only partial epithelial loss and no dermal exposure.
  - contusions disrupt the subepithelial tissue vascular supply:
    - ♦ predisposes to anoxic necrosis, secondary loss of epithelium and bacterial colonisation.
  - less dramatic in appearance and most heal without further intervention.
  - may present with delayed and severe tissue damage (e.g. rope burns):
    - ♦ can mislead the clinician in their initial assessment of the injury:
      - adequate follow-up essential.

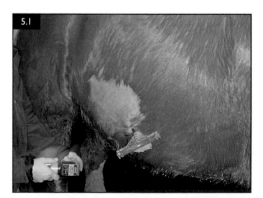

FIG. 5.1 A puncture wound to the right abdominal wall caused by a metal stake used to carry an electric fence wire. The object had penetrated the wall and damaged a loop of small intestine.

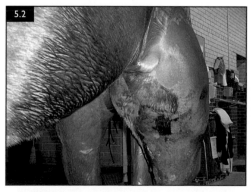

FIG. 5.2 Swollen thigh as a result of a contusion and hematoma. Note the area of alopecia in the craniomedial aspect of the limb. Hair loss is one of the early signs of tissue necrosis prior to sloughing, usually seen 3–5 days after the original injury.

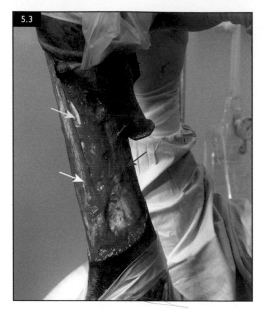

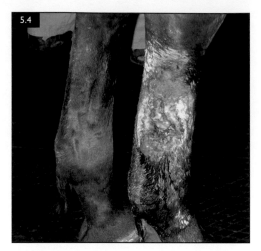

FIG. 5.4 issue sloughing on the lateral aspect of the metacarpal area subsequent to a rope burn sustained 6 days previously.

FIG. 5.3 Acute degloving injury of the hind cannon following the leg being trapped through the floor of a horse trailer. The long digital extensor tendon is exposed (white arrow), along with the flexor tendons (red arrow), and areas of the third metatarsus (yellow arrow).

♦ may develop septic cellulitis with tissue damage, necrosis and sloughing within first 5–7 days (Fig. 5.4).

Stages of wound healing

- after wounding a series of events follows that constitute the healing process.
- timing and cellular elements of this process characterise the different stages in each separate tissue.
- stages overlap each other and occur in programmed succession to achieve a final result.
- practitioner should modulate these stages, through intervention, to:
  ○ facilitate and speed the healing process.
  ○ promote a quick return to proper function.
  ○ improve the cosmetic outcome.
- five stages of wound healing are: haemostasis, inflammation, debridement, repair and maturation.

Haemostasis

- may last a few minutes to a few hours.

- extravasation of vascular elements and exposure of subendothelial structures leads to:
  ○ aggregation of platelets, which contributes to clot formation.
  ○ platelets also responsible for fibrin mesh and chemoattraction of neutrophils.

Inflammation

- acute inflammatory phase starts within minutes of a wound occurring:
  ○ lasts a few hours to days depending on severity of injury and degree of contamination.
- characterised by the presence of neutrophils (bactericidal):
  ○ protect the body against invasion by foreign organisms or substances.
  ○ release of local chemotactic factors acts as migration signal for the neutrophils.
  ○ other local factors responsible for:
    ♦ vasodilatory response and an increase in vascular permeability.
  ○ activation and death of neutrophils releases enzymes and inflammatory mediators.
  ○ neutrophils are not necessary for wound healing in the absence of infection.
- balance between acute and chronic inflammation for a wound to heal optimally by second intention:

5

- o acute inflammation has beneficial effects on wound healing – encouraged.
- o chronic inflammation promotes the production of exuberant granulation tissue – discouraged.

Debridement

- begins within a few hours of the injury and duration depends on severity of the injury.
- enzymatic breakdown of debris by substances released from dead neutrophils.
- Monocytes are chemoattracted into the wound at the same time as neutrophils:
  - o responsible for elimination of debris, necrotic tissue, foreign substances and microorganisms.
  - o necessary for wound healing to progress.
  - o transform into macrophages once in the wound.
  - o remain there for days to weeks, regulating the progression of wound healing.

Repair

- characterised by the presence of fibroblasts:
  - o produce collagen and connective tissue matrix (granulation tissue).
  - o lead to wound contraction.
- granulation tissue scaffold facilitates the migration of epithelial cells.
- Second-intention or open wound healing (Fig. 5.5):
  - o epithelialisation and contraction are ultimately responsible for wound repair.
  - o only possible if the wound is clear of:
    - ♦ debris, infection, tissue necrosis and exuberant granulation tissue.
  - o epithelialisation begins as soon as 12 hours post-wounding in primary healing:
    - ♦ after 4 or 5 days in secondary wound healing.
  - o wound contraction is important in secondary wound closure and stops with:
    - ♦ contact inhibition between cells.
    - ♦ excessive tension.

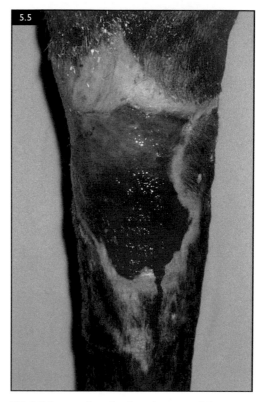

FIG. 5.5 A wound on the dorsal aspect of the upper cannon that is being treated by second-intention healing. Note the healthy granulation-tissue bed and active epithelial edges.

- ♦ exuberant granulation tissue.
- ♦ full-thickness grafts that are applied before the 5th day of healing.

Maturation

- return to normal structure and function of the connective tissue is characterised by:
  - o balance between collagen synthesis and lysis.
  - o return of normal architecture with adequate pattern of collagen fibre orientation.
  - o leads to increased wound tensile strength (approx. 80% of original tissue).
- after the maturation phase, the tissue evolves through:
  - o subclinical period of remodelling, which may last weeks to months.

Factors affecting wound healing

Size of horse (horses versus ponies <1.48 m)

- ponies heal faster than horses:
  - their acute inflammatory phase is of shorter duration.
  - their leukocytes have a different capacity to aid with wound healing.
- healing phases of pony wounds are usually much shorter than in horses.
- their capacity to develop exuberant granulation tissue is significantly reduced.

Use of NSAIDs

- NSAIDs shown to have adverse effects on wound healing and graft survival:
  - reduce acute inflammation that is bactericidal.
  - responsible for early wound contraction and epithelialisation.
- use during this stage of wound healing should therefore be limited:
  - lowest effective dose.
  - alternative analgesics should be considered:
    - opioids alone (e.g. butorphanol 0.02–0.1 mg/kg i/m or i/v q4–6 h).
    - combinations (e.g. ketamine 0.5–1 mg/kg and methadone 0.05 mg/kg i/m, q4–8 h).

Nutritional status

- food deprivation and malnutrition can have a deleterious effect on wound healing.
- conditions associated with a reduction in plasma protein (e.g. malnutrition):
  - delay repair phase of wound healing by preventing the onset of fibroplasia and diminishing the tensile strength of a wound.

Hypovolaemia, hypotension and hypoxia

- sudden hypovolaemia due to considerable blood loss delays healing:
  - impairs the microcirculation due to reduction in circulating blood volume.
  - leads to local tissue hypoxia and reduced micronutrient delivery.

- local hypovolaemia can occur following strangulating trauma:
  - smooth-wire injuries, overzealous use of pressure bandages, or rope wounds.
  - prevent or reduce vascular supply to the distal extremity.
  - resulting hypoxic tissue damage.
  - possible necrosis leading to tissue oedema and sloughing.

Hypothermia

- optimal wound healing occurs at environmental temperatures of around 30°C (86°F).
- effect of temperature changes on wound healing in horses has not been studied:
  - except in extreme temperature conditions, it is thought that temperature change should have no net effect on the process of wound healing.

Infection

- wounds not promptly decontaminated and debrided:
  - severe wound contamination, lack of vascular supply and the presence of foreign material may lead to wound infection.
  - infection interferes with and delays wound healing.
  - severity of wound contamination is generally directly related to its proximity to the ground.
- contamination is not synonymous with infection:
  - bacterial colonisation is necessary for the latter to occur.
- infection is an uncommon occurrence in a properly treated wound:
  - largest challenge is when bone or synovial structures are involved, or major tissue necrosis is present.
  - bacteria may become quickly established, with catastrophic consequences.
- debridement of a wound is an important component of wound treatment:
  - eliminates necrotic tissue, foreign material, blood and exudates.
  - helps reduce bacterial load to levels that the host immune system can deal with.

- antibiotics will not prevent an infection in the presence of other predisposing factors:
  - use of systemic or topical antibiotics will depend on the assessment of the wound.

Topical medications

- used for:
  - preventing excessive granulation tissue.
  - speeding up healing.
  - improving the cosmetic outcome.
  - preventing wound infection.
- no single product effective in all these areas:
  - lack of proper clinical trials showing data supporting the use of a single product.
- existing medications include antimicrobials, antiseptics, irritants and cell-function modulators:
  - antimicrobials and antiseptics may be indicated in the initial stages of wound healing to reduce the bacterial load.
  - use of antimicrobials is not indicated once a granulation-tissue bed is present.
  - corticosteroids are used to prevent excessive granulation-tissue production:
    - **dose-dependent effects on healing – delayed with higher doses.**
    - **use cautiously.**
  - local anaesthetics close to a wound edge should be avoided (toxic for leukocytes):
    - either:
      - perform regionally by perineural infiltration.
      - inject local anaesthetic as far from the wound edge as possible.
      - do not to use any vasoconstricting agents (e.g. epinephrine).

Movement

- excessive movement promotes exuberant granulation tissue and interrupts healing:
  - particularly where poor local musculature, such as distal limbs (Fig. 5.6).
- movement should be controlled using bandages, splints or casts.

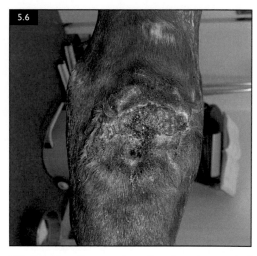

FIG. 5.6 Chronic wound on the dorsal aspect of the hock and upper cannon bone. This had been treated for 9 months, but because of excessive movement and difficulty in maintaining bandages in place, the result was very poor, with thickening of the skin surrounding a non-healing central area.

  - cast immobilisation is a useful technique in distal limb wounds.

Necrotic tissue

- delays or prevents wound healing by:
  - preventing cell proliferation and adequate vascular supply.
  - potentiates ongoing inflammation and wound infection.
  - debridement of necrotic tissue can substantially improve wound healing.

Tumours

- presence of neoplastic tissue prevents wound healing by:
  - impairing cell function and normal contraction and epithelialisation.
- open wound healing can lead to metaplasia of tissues:
  - predisposing the site to fibroblastic sarcoid.

Types of closure

Primary closure (Fig. 5.7)

- apposition of the skin epithelium, with restoration of the skin surface:
  - mechanism by which surgical wounds heal.

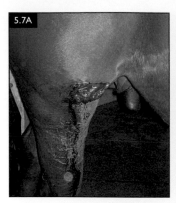
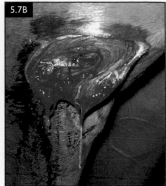
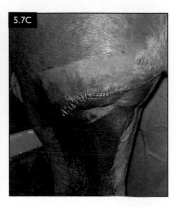

FIG. 5.7 An acute laceration of the right stifle region of a Thoroughbred racehorse. (A) Wound has been lavaged, explored and cleaned prior to surgical debridement. (B) Wound has been sharply debrided and lavaged prior to repair. (C) Repaired in anatomic layers, including muscle, fascia and subcuticular layers, before skin staples are used to close the skin.

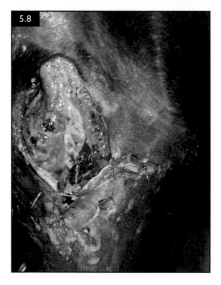

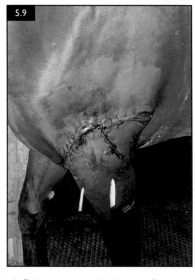

FIG. 5.8 Wound dehiscence as a result of excessive tension, motion and tissue necrosis (96 hours after primary closure).

FIG. 5.9 Primary wound closure after an acute laceration. Note the everting suture pattern used (horizontal mattress) and the placement of Penrose drains.

- chosen when:
  - adequate vascular supply.
  - minimal tension across the sutured wound.
  - none or minimal bacterial contamination.
  - if these criteria are not met, dehiscence will occur (Fig. 5.8).
- partial primary closure is an option for wounds that may have:
  - area of isolated tissue necrosis.
  - excessive skin loss.
  - require wound drainage in cases of excessive dead space or contamination.
- when drainage is required:
  - most dependent part of wound should not be sutured.
  - use of drains should be considered (Fig. 5.9).

Delayed primary closure

- performed prior to the onset of fibroplasia, approximately 3–5 days after wounding.

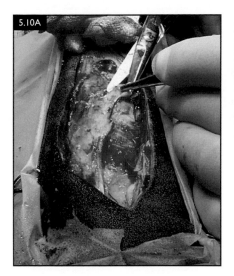

FIG. 5.10 This horse sustained a laceration to the dorsal mid-cannon bone due to entanglement with a wire fence. It was elected to treat the wound by delayed primary closure with a regimen of daily wound lavage and cleaning, topical hydrogel application, compression bandaging and systemic antibiotics and NSAIDs. **(A)** 5 days post-injury the wound was debrided under general anaesthesia prior to lavage and surgical repair. **(B)** The wound edges were mobilised by undermining and closed with a combination of subcuticular sutures, skin staples and vertical mattress bolster monofilament nylon sutures.

- allows wound to drain and clear excessive debris or bacterial load before being closed:
  - decreases the chance of dehiscence (Fig. 5.10).
  - improves vascular supply and debridement.
  - decreased healing time when compared with second-intention wound healing.
- increases initial wound-edge retraction:
  - may increase wound tension and make apposition more difficult.
- meticulous wound care prior to wound closure is essential:
  - protect and debride the wound and remove exudates and necrotic debris.
  - prevent wound desiccation and cross-contamination.
  - confine the animal.
  - use permeable bandages applied daily either as:
    - dry-to-dry or wet-to-dry modalities to debride and support the wound.
  - wound lavage.
  - careful use of systemic antibiotics and anti-inflammatories to control infection and excessive inflammatory reaction in some cases.

Secondary or open wound closure

- first-intention healing is not possible:
  - excessive motion, tension, tissue damage or infection (Fig 5.11).
- complications include:
  - development of exuberant granulation tissue in lower extremities of the limb.
  - time required for re-epithelialisation (Fig. 5.12).
- wound management during healing is time-consuming and expensive.
- relies on formation of an appropriate granulation-tissue bed for the wound to contract and re-epithelialise:
  - starts to form as early as 72 hours and proceeds rapidly if the wound conditions are adequate (i.e. lack of infection and immobility).
  - clinician must manage the wound to facilitate wound contraction and epithelialisation by maintaining an appropriate wound environment.

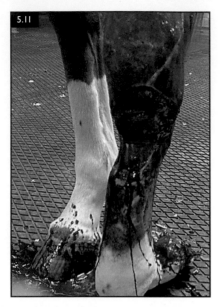

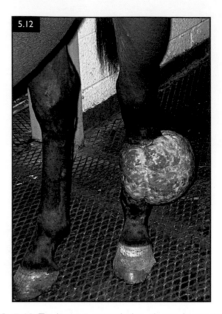

FIG. 5.11 An extensive wound to the dorsal aspect of the tarsal and proximal metatarsal regions, typically encountered in barbed-wire fence injuries. Due to its size and location, this wound is not amenable to primary closure. Secondary closure, or open wound healing, was selected.

FIG. 5.12 Exuberant granulation tissue is present after a wound in the dorsolateral aspect of the metatarsal region of a 9-year-old Quarter horse.

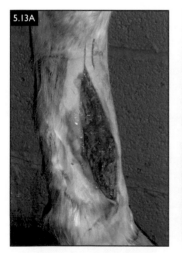

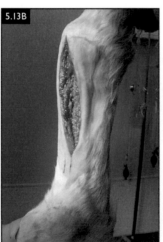

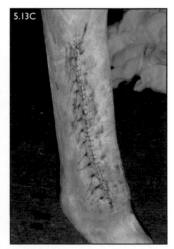

FIG. 5.13 This horse sustained a wound of the medial cannon region 8 days previously (A). The wound was cleaned, lavaged, and bandaged every day, and systemic antibiotics and NSAIDs were given (B). The wound was repaired by delayed secondary closure (C) and healed without incident.

Delayed secondary closure

- performed by suturing a wound after the process of fibroplasia has produced a healthy granulation-tissue bed.

- correct application can:
  - prevent development of exuberant granulation tissue.
  - considerably accelerate the horse's recovery.

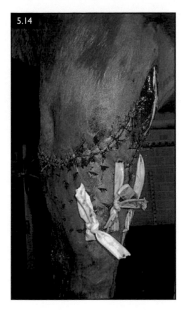

FIG. 5.14 The 'mesh-expansion' technique. Note the direction of the stab incisions parallel to the direction of the wound.

- repair should involve the following (Fig. 5.13):
  - ○ debridement of superficial layer of granulation-tissue bed.
  - ○ skin edges undermined and freshened.
  - ○ apposition with minimal tension.
- additional tension-release techniques, such as mesh expansion (Fig. 5.14):
  - ○ decrease the likelihood of dehiscence of the suture line.

Wound closure

Preparation, lavage and debridement

- assess and treat all wounds in a quiet, clean and well-lit environment if possible.
- examine the whole horse thoroughly:
  - ○ prior to any drug administration, assess for signs of shock and severe blood loss:
    - ♦ tachycardia, pale/muddy mucous membranes, prolonged capillary refill time (CRT), skin tent.
- sedation may be necessary in many cases to allow safe and effective treatment.
- interaction with any wound must be carried out with clean hands or preferably gloved.

- evaluation of a wound should include:
  - ○ move the horse, if necessary, to a better environment.
  - ○ initial assessment of the wound.
  - ○ apply sterile hydrosoluble gel to the wound surface.
  - ○ clip hair from the surrounding area to at least 5 cm from the wound edges.
  - ○ lavage wound with sterile saline (hose with cold water if not available):
    - ♦ manually (using gloves +/– gauzes) remove all gross contamination:
      - – dirt, straw, hair and other foreign material.
  - ○ apply additional sterile hydrosoluble gel to wound:
    - ♦ scrub wound periphery with a suitable solution (chlorhexidine soap):
      - – avoid contamination of the wound bed.
  - ○ irrigate the wound with 1 litre of 0.1% povidone–iodine solution either with:
    - ♦ Waterpik™ or a 60 ml syringe through a 19-gauge needle.
    - ♦ dislodge contamination (+/– sterile gauzes) without driving fluid into wound.
  - ○ explore the wound in a sterile manner with:
    - ♦ sterile metal probe or sterile gloved digit.
    - ♦ assess presence and depth of wound tracts.
    - ♦ determine:
      - – anatomic structures involved.
      - – tissue deficits.
      - – vascular supply.
      - – dead space.
      - – level of deep contamination and potential deep tracts.
- develop therapeutic plan and decide whether primary or secondary closure is necessary.
- surgical debridement is essential:
  - ○ performed with careful resection of necrotic tissue.
  - ○ preserve all viable tissue and vital structures.
  - ○ over-debridement may delay the healing process:
    - ♦ questionable skin flaps maintained for a few days ('biological' dressing):

TABLE 5.1 Types of suture material available to the equine clinician

SUTURE MATERIAL	ABSORBABLE	CONFORMATION	COMMERCIAL NAME
Polydioxanone	Yes	Monofilament	PDS/Biosyn
Polypropylene	No	Monofilament	Prolene
Polyglactin 910	Yes	Multifilament	Vicryl/Polysorb
Polyglycolic acid	Yes	Multifilament	Dexon
Polymerised caprolactam	No	Multifilament	Supramid
Poliglecaprone 25	Yes	Monofilament	Monocryl
Polyester	No	Multifilament	Ethibond
Polyglyconate	Yes	Monofilament	Maxon

  - debrided once fibroplasia has started.
    ○ debridement can be a staged procedure and continue for several days.
- debridement can be accomplished by medicinal maggot therapy:
    ○ safe, effective and controlled method for chronic wounds.
    ○ maggots are sterile or disinfected.
    ○ preference to feed on non-vitalised tissue, purulent exudate and metabolic wastes of a wound.
    ○ produce disinfection, debridement, stimulation of healing, and inhibition and eradication of biofilms.
    ○ main use is for the treatment of recalcitrant foot abscesses.
- hydrosurgery systems (Versajet™, Smith & Nephew, St Petersburg, USA) (Fig. 5.15) are available to perform precise wound debridement.

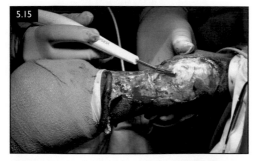

FIG. 5.15 Use of the Versajet system to debride an extensive laceration of the dorsal pastern and fetlock.

Suture materials and patterns

- choose the appropriate suture material, needle and pattern.
- suture materials can be classified as:
    ○ absorbable/non-absorbable and mono- or multifilament (Table 5.1).
- suture materials should be:
    ○ user friendly.
    ○ prevent a large foreign-body reaction.
    ○ maintain tension and knot security until the tissue has healed.
    ○ minimise the risks of harbouring bacteria.
- monofilament absorbable material (polydioxanone) is ideal in subcutaneous tissues.
- non-absorbable material (polypropylene) ideal in the skin.
- size of needle and suture material should be proportional to the thickness of the tissue:
    ○ subcutaneous and internal tissues, a round-tipped non-cutting needle is preferable.
    ○ reverse cutting needle is indicated in the skin.
    ○ suture size 2/0 or 3/0 (3/3.5 metric) appropriate for subcutaneous tissues and skin.
    ○ large amount of tension is expected:
        ♦ larger suture size or use of a different suture pattern is required.
- suture pattern chosen should be adequate to maintain apposition of the tissues:
    ○ tension not a concern – simple or cruciate mattress interrupted pattern.

5

- tension is anticipated – use a pattern that provides tension relief:
  - near–far–far–near or a vertical mattress.
- appositional or mildly everting suture patterns are adequate in the skin.
- cosmetic concerns:
  - subcuticular simple continuous pattern if no concerns about tension.
- closure of dead space will prevent development of seromas and speed up healing:
  - suture subcutaneous layers.

Wound drainage

- removes excess fluid generated in a wound:
  - minimise anatomic dead space.
  - promote healing.
  - decrease bacterial proliferation.
  - reduce the risk of infection.
- two types of draining methods:
  - **Passive drains:**
    - use gravity to facilitate fluid elimination.
    - strategically placed incisions, ventrally located relative to the drainage area:
      - gentle body motion helps passive drainage.
  - devices such as Penrose drains made of latex or sterile tubing:
    - not exiting directly through the wound (Fig. 5.16).
  - main disadvantage is possibility of an ascending infection:
    - place for short period of time and where absolutely necessary.
  - use in cases of synovial wounds or osteosynthesis is controversial.
  - **Active drains:**
    - requires suction system that provides a continuous negative pressure:
      - removes exudate as soon as produced.
      - prevents retrograde contamination of the drained area.
    - manufactured with a large syringe, tubing and a needle (Fig. 5.17):
      - commercially available.
    - careful emptying/reconnection necessary to maintain the closed system.
    - strategic location must be chosen for attaching the suction system.
    - presence of drain tubing in the wound may stimulate foreign-body reaction and fluid production:
      - drain should be removed immediately.
      - never in place for more than 72 hours.

FIG. 5.16 A Penrose drain is shown placed away from the wound edges and into an area of dead space to facilitate wound drainage and prevent the formation of a seroma.

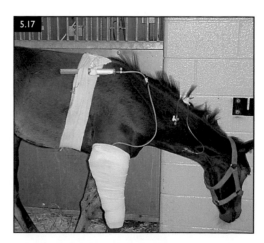

FIG. 5.17 An 'active' drain, following a fracture repair. Note the location of the collection chamber of the drain and its simple construction.

Postoperative care

Dressings and immobilisation
(Tables 5.2, 5.3)

- Bandages are composed of three layers:
  - contact.
  - absorbent or intermediate.
  - protective or shell.
- **Contact:**
  - immediate layer on the wound surface.
  - fulfils all or some of the functions of protection, debridement, absorption and occlusion, and a vehicle for topical medications.
  - absorption capacity of contact and intermediate layers determined by their hydrophilic properties.
  - depending on their permeability to liquid (exudates) and gas (oxygen), they are:
    - ◆ **Occlusive dressings:**
      - exudates are not allowed to evacuate the surface.
      - maintain a high level of moisture in the wound.
      - may promote additional exudation, exuberant granulation tissue and inflammatory response.
      - results in delayed healing.
      - **should not be used in primary post-surgical wounds.**
      - facilitate wound dehiscence by maintaining an increased level of moisture at the wound site.
      - best applied once the quantity of exudates is reduced and wound starting to granulate.
      - use in equine wound healing is still controversial.
    - ◆ **Semi-occlusive:**
      - more versatile.
      - more adjustable, particularly during open wound management.
      - use of a non-adherent, semi-occlusive dressing limited to:
        - ◇ primary wound healing.
        - ◇ secondary wound healing once granulation-tissue is present.

- during exudative stages of secondary wound healing:
  - ◇ consider use of semi-occlusive bandages.
  - ◇ dry-to-dry, wet-to-dry and wet-to-wet fashion (see Table 5.3).
  - classification of bandages according to moisture gradients:
    - ◆ bandages can be used with varying absorptive capacity:
      - regulate the level of moisture at the wound surface.
    - ◆ dry-to-dry (dry contact layer, dry intermediate layer):
      - used during the first 48–72 hours when debridement is needed.
    - ◆ wet-to-dry (wet contact, dry intermediate):
      - increased wound exudation in inflammatory/debridement phases.
      - applied until a granulation-tissue bed is visible.
    - ◆ wet-to-wet (wet contact, wet intermediate).
      - used if excessive exudation.
- Absorbent or intermediate
- Protective or shell.
- primary wound healing (surgical incision closed with minimal tension and good apposition):
  - bandages have primarily a protective role.
  - non-adhesive dressings such as telfa or melolite.
  - well-placed bandage.

Topical medications

- application in initial stages of wound healing rarely necessary if:
  - adequate wound debridement, environment and motion control.
- management of exuberant granulation tissue is discussed elsewhere (see page 164).
- recent developments which can be used topically to enhance wound healing:
  - platelet-rich plasma (PRP)
  - Manuka (*Leptospermum scoparium*) honey.

TABLE 5.2 Types of dressings and their applications according to their moisture content		
TOPICAL DRESSING	**COMMENTS**	**INDICATIONS**
Corticosteroids	Once daily apply a clear thin layer. Prevent granulation tissue, speed healing, may decrease fibroplasia and delay wound contraction and epithelialisation, perhaps dose related.	Prevents exuberant granulation tissue. Use in repair phase.
2.5% ketanserin	Once daily apply a clear thin layer. (NB: use gloves.) Prevents granulation tissue. No bandage needed. Experimental results equivocal; do not compare with bandaging management.	Prevents exuberant granulation tissue.
Silver sulfadiazine (1%)	Once daily apply a clear thin layer. No effect on wound healing. Good effects as antimicrobial *in vitro*.	Infected wounds, burns.
Manuka honey	Once daily apply a clear thin layer (place in fridge if too runny). Antimicrobial. Positive effects on healing. 30 ml/100 cm^2	Use throughout initial stages of wound healing until granulation tissue is covering wound.
Platelet-rich plasma	Twice, 7–10 days apart on a gel preparation.	Use until granulation tissue is covering the wound bed.
Biological dressings	No advantages in bacterial proliferation, inflammatory reaction, wound contraction or epithelialisation. Some dressings are occlusive.	Use during repair phase.
Amniotic membrane	Decreases granulation tissue; faster healing. Time-consuming to produce; costly.	Use during repair phase.
Hydrogels	Humans: pain relief, inhibition of bacterial proliferation, debrides devitalised tissue, provides moist healing environment. Horses: no effect on contraction and epithelialisation shown. Costly.	Use in the late stages of the repair phase.
Colloids	Occlusive.	Use in wounds with established granulation tissue, advanced contraction and decreased fluid production.
Nitrofurazone	Once daily apply a clear thin layer. Helps decrease bacterial load. Non-adherent. Conflicting results as to its effects on contraction and epithelialisation.	Use when granulation tissue is present and a small amount of exudate, which would indicate a larger bacterial load.
Tri-peptide copper complex	Rats: benefits chronic ischemic wounds and accelerates contraction and epithelialisation.	No horse studies are available.
Occlusive dressing	Horses: delays healing, increases inflammatory reaction, increases exudate production.	Use in the late stages of the repair phase.
Gauze	Semi-occlusive. Dry-to-dry or wet-to-dry.	Debriding wounds. Non-adhesive. Maintains wound moisture while allowing drainage.
Telfa, melolite	Semi-occlusive.	Non-adhesive. Use after granulation tissue present.
Tegaderm, Opsite	Occlusive.	Increases wound moisture.
Vigilon, Nu-Gel, 2nd skin	Occlusive, hydrogel.	Increases wound moisture. Promotes granulation tissue.
Tegasorb, Duoderm, Intrasite	Occlusive, hydrocolloid.	Become incorporated into the wound. Promote epithelialisation and granulation.

TABLE 5.3 Types of semi-occlusive bandages

BANDAGE	CONTACT LAYER	INTERMEDIATE LAYER	USES
Dry-to-dry	Dry gauze, permeable	Dry, hydrophilic, permeable	Initial stages (debridement) of wound healing when mechanical debridement is needed
Wet-to-dry	Wet gauze, permeable	Dry, hydrophilic, permeable	In wounds with moderately thick and abundant exudate
Wet-to-wet	Wet gauze, permeable	Wet, hydrophilic, permeable	In wounds with a large amount of thick exudate

Topical negative pressure (TNP)

- use of a vacuum-assisted device in the treatment of wounds by second-intention closure (Fig. 5.18):
  - ○ possible effects include increased blood flow, promotion of angiogenesis and induction of cell proliferation.

Wound support

- wound immobility is a crucial factor in healing:
  - ○ excessive motion promotes dehiscence and delays wound healing.
  - ○ promotes exuberant granulation tissue.

- ♦ inadequate wound-motion management is common in clinical practice.
- ○ different degrees of immobilisation can be accomplished with:
  - ♦ appropriately applied bandage +/– splint or cast (Fig. 5.19).
  - ♦ heavily layered bandage or modified Robert Jones bandage provides:
    - – immobility directly related to the number of layers applied.
- ○ important to maintain conformation and uniform pressure of the bandage to avoid complications, such as pressure sores and undue wound irritation (Fig. 5.20).
- Splints:
  - ○ light
  - ○ easy to contour
  - ○ waterproof.
  - ○ strong.

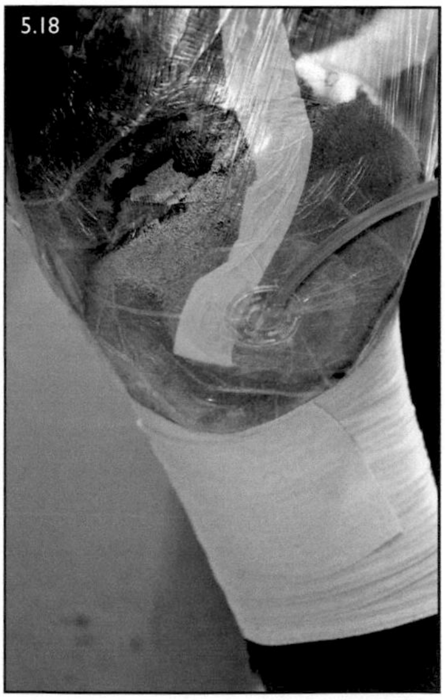

FIG. 5.18 Vacuum-assisted device on a granulating wound of the tarsus covered in an occlusive dressing.

FIG. 5.19 Use of a foot cast for the treatment of a severe heel bulb injury.

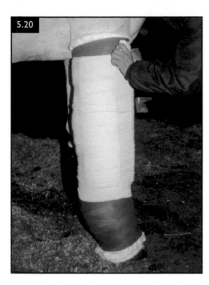

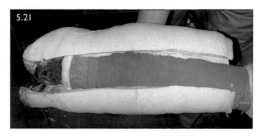

FIG. 5.21 A bivalved cast or cast–bandage combination is ideal to treat injuries that require immobilisation and daily care. Note that the limb is not allowed to bear weight during wound therapy.

FIG. 5.20 This horse has a wound on the dorsal aspect of the carpus, which has been surgically repaired. The limb is bandaged post-surgery in a full-limb Robert Jones bandage, with splints fitted to the lateral and dorsal aspects of the limb from just distal to the elbow to just above the fetlock.

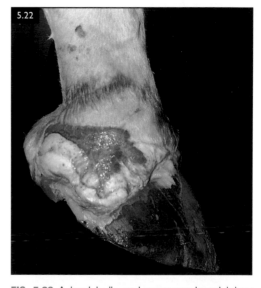

FIG. 5.22 A heel bulb and coronary band injury treated by surgical debridement, lavage and stabilisation in a foot cast. This is at the first change of the cast and already the wound is healthy, healing well by second intention and showing no excessive granulation tissue.

- ○ cheap   ○ without sharp edges.
- ○ PVC piping is ideal.
- ○ allows access to wound as required.
- ○ customise to appropriate length and width for individual horse.
- ○ inappropriately placed or padded splints:
  - ♦ may produce large pressure sores, particularly in foals:
    - – too little padding can lead to pressure sores.
    - – too much padding will allow the splint to move around the horse's limb.
- • application of a cast may be preferable for wounds where daily access is not required.
- • bivalved cast or bandage–cast combination permits the best of both worlds:
  - ○ regular access to wound.
  - ○ provides very adequate immobilisation (Fig. 5.21).
  - ○ useful for wounds affecting synovial cavities, particularly:
    - ♦ tendon lacerations involving the digital flexor tendon sheath.
- ○ casts can be useful for the treatment of open wounds, especially of the distal limb:
  - ♦ facilitate wound healing.
  - ♦ prevent formation of exuberant granulation tissue (Fig. 5.22).
- ○ daily evaluation for indications of pressure sores:
  - ♦ cracks.
  - ♦ excessive exudates.
  - ♦ inflammation at the top of the cast.
  - ♦ hot spots.
  - ♦ lack of use by the horse.

- o any doubts pertaining to complications:
  - ◆ remove the cast and evaluate the limb.
  - ◆ pressure sores in areas such as proximal sesamoid or accessory carpal bones:
    - – may potentially lead to serious and expensive complications.

Exuberant granulation tissue (EGT) ('proud flesh') management

- stops the process of wound contraction and epithelialisation.
- common complication of second-intention wound healing of distal extremity wounds.
- occurs with:
  - o individual predisposition.
  - o conditions of motion and infection.
- results in an undesirable cosmetic and functional outcome.
- higher prevalence in horses than ponies (height <1.48 m):
  - o differences in inflammatory process, cellular activity and activation of growth factors.

FIG. 5.23 Trimming granulation tissue. The blade is maintained parallel to the wound surface and the granulation tissue is trimmed from distal to proximal. Note the amount of blood present during this procedure.

- specific cause(s) are unknown.
- dealing with large amounts of EGT can be frustrating, lengthy and costly.
- prevention of motion and infection early in wound healing will limit incidence.
- once identified, its management requires aggressive therapy.
- no treatments consistently prevent or eliminate EGT in all horses.
- use of caustics and irritants has not been shown to be effective:
  - o wound irritation may further delay wound healing.
- topical corticosteroids can be beneficial:
  - o response is dose dependent.
  - o large doses can have undesirable side effects.
- principles of treatment (Fig. 5.23) are:
  - o radical surgical excision.
  - o topical use of corticosteroids.
  - o reduction of motion.
  - o elimination of wound infection.
  - o use of grafting techniques.

Wound complications

- formation of exuberant granulation tissue.
- dehiscence.
- seroma formation.
- infection.
- all can be related, and the common result is failure to heal.

Wound dehiscence

- may be due to:
  - o wound infection
  - o inadequate blood supply
  - o excessive motion.
  - o inappropriate holding power of the suture line:
    - ◆ either tissue- or suture-related.
  - o identification of the specific problem is essential before effective treatment.
- infection-related wound dehiscence may take several days to occur:
  - o treat the underlying infection for a few days.
  - o freshen wound edges prior to closure.
  - o re-suture dehisced wound edges if healthy and strong.
- tension-related dehiscence may occur within the first 24 hours:
  - o different closure techniques should be used.

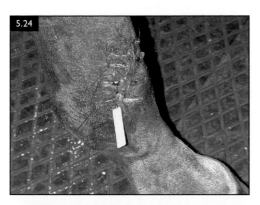

FIG. 5.24 The use of bolster sutures. Polyethylene tubing is used to distribute suture tension and minimise the chances of suture-induced tissue necrosis.

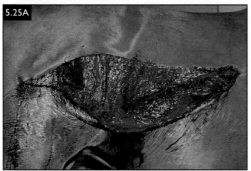

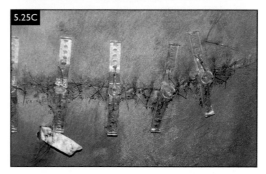

FIG. 5.25 These three illustrations show how tissue approximators can be used in cases of large separation of skin edges. **(a)** A flank wound prior to repair. **(b)** The same wound repaired with the use of tissue approximators. **(c)** Detail of the tissue approximators. The 'dial' piece in the centre is used on a daily basis to approximate the wound edges. (Photos courtesy D Trout.)

- ○ surgical techniques to reduce tension include:
  - ◆ mattress suture patterns
  - ◆ bolster sutures (Fig. 5.24).
  - ◆ mesh-expansion techniques.
  - ◆ stents.
  - ◆ plastic reconstruction techniques.
  - ◆ tension-relief sutures.
  - ◆ Wound edge approximators (Fig. 5.25):
    - – allows daily approximation of the wound edges.
    - – prevents retraction of the wound.
    - – evenly spreads the tension of each individual suture.
    - – may require several days before apposition of wound edges.

Seroma formation

- • usually occurs due to:
  - ○ increased subcutaneous dead space
  - ○ inadequate haemostasis.
  - ○ presence of foreign bodies
  - ○ severe inflammation.
- • small/innocuous seromas should be monitored and left alone to resolve with time:
  - ○ hot compresses can be applied to help resolution.
- • large or growing seromas jeopardise the suture line:
  - ○ prior to intervention, evaluate seroma cavity and contents by ultrasonography:

TABLE 5.4 Advantages and disadvantages of the different types of skin grafts

TYPE OF GRAFT	ADVANTAGES	DISADVANTAGES
Full-thickness grafts (pinch, punch, tunnel or sheet)	• Practical in the standing horse • No specialised equipment required • Better cosmetic outcome	• Morbidity of donor site • Slower revascularisation and 'graft-take' • Slower healing • Limited graft size • Greater primary contraction
Split-thickness grafts (thin, medium or thick sheet grafts)	• Quicker revascularisation and 'graft-take' • Faster healing • Abundance of donor skin	• Specialised equipment required • General anaesthesia required • Poorer cosmetic outcome • More secondary contraction

- ♦ information on size of seroma.
- ♦ identify any foreign bodies.
- ♦ volume and nature of the contents.
- ○ drained by aspiration or by providing a drainage route.
- ○ careless aspiration may introduce bacteria and severe infection.
- ○ may only provide temporary relief:
  - ♦ seroma reforms, usually in 24–48 hours.
- ○ drain placement and pressure bandage is indicated in large seromas.

Skin grafting

Overview

- purposes of skin grafting are:
  - ○ accelerate and facilitate the process of wound healing.
  - ○ prevent development of exuberant granulation tissue.
  - ○ improve cosmetic outcome.
- full-thickness grafts include entire dermis and epidermis.
- partial- or split-thickness grafts include entire epidermis and variable portion of dermis:
  - ○ split-thickness grafts are 0.5–0.76 mm thick.
  - ○ harvested with dermatome (Fig. 5.26):
    - ♦ ensures consistent and uniform graft thickness.
- each type of graft has definite advantages and disadvantages (Table 5.4).
- healing process of a skin graft involves:
  - ○ adherence and graft nutrition (48 hours).
  - ○ revascularisation (2–4 days).
  - ○ firm union (7–10 days).

- ○ contraction (primary or immediate, secondary or delayed).
- graft failure is usually due to:
  - ○ infection impeding vascular supply at the microvascular level:
    - ♦ increased fibrinolytic substances, which compromise graft adherence.
  - ○ excessive motion affects graft attachment and revascularisation:
    - ♦ prevented by appropriate bandaging techniques or cast immobilisation.
    - ♦ particularly important in high-motion areas such as joints.
  - ○ accumulation of fluid beneath the graft may prevent revascularisation:
    - ♦ physical separation of donor and recipient elements.
    - ♦ leads to graft ischemia and necrosis.
- meticulous preparation of the donor graft and recipient sites is essential:

FIG. 5.26 Drum dermatome is shown during split-thickness graft collection. (Photo courtesy S Barber.)

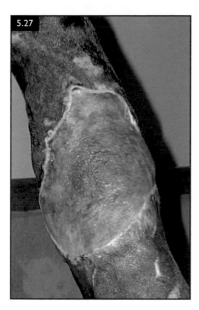

FIG. 5.27 A granulation-tissue bed ready for grafting must be flush with the epithelial edges and appear well vascularised, with minimal exudates and no discoloured tissues.

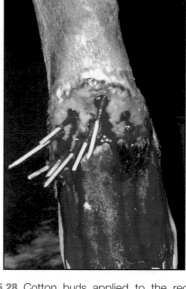

FIG. 5.28 Cotton buds applied to the recipient site of punch grafts to facilitate haemostasis and improve graft 'take'.

○ recipient site preparation within 24 hours of grafting.
○ healthy granulation-tissue bed characterised by:
  ♦ good vascular supply.
  ♦ uniform granulation tissue with absence of cracks or fissures.
  ♦ level with peripheral epithelium (Fig. 5.27).
○ absence of large bacterial load or suppurating focus is mandatory:
  ♦ trim covering of healthy granulation-tissue bed:
    – eliminates surface contaminants.
    – level with peripheral epithelium.
    – bleeds profusely when trimmed (see Fig. 5.23):
      ◊ avoid immediately prior to grafting.
  ♦ after trimming, bandage wound in a sterile manner:
    – facilitates haemostasis.
  ♦ topical preparations (e.g. silver sulfadiazine or diluted Dakin's solution):
    – helps ensure low or absent bacterial load prior to grafting.

FIG. 5.29 Punch grafts in place. Note that the grafts are slightly below the granulation-tissue surface and are separated by about 0.5 cm.

• proper grafting procedures are vital to achieve success:
  ○ accurate haemostasis while placing the graft (Figs. 5.28, 5.29).

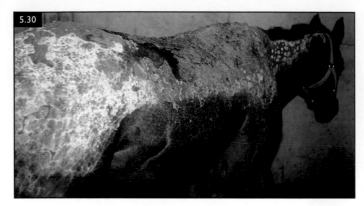

FIG. 5.30 This horse was involved in a barn fire and sustained severe burns over most of the dorsum of its body. At this stage the horse has had several sessions of pinch grafting and the area affected by the burns is gradually healing by second intention.

Burns

- most frequently due to barn or forest fires (Fig. 5.30).
- most important aspect when evaluating burns in the horse are:
  - provide analgesia and quickly evaluate the damage.
  - prognosis is directly related to the extent and severity of the burns:
    - amount involved expressed as a percentage of total body surface area:
      - severe burns (50% body surface area) have a high mortality.
    - poor prognosis is an indicator for euthanasia.
- severity of burns has been classified from 1st to 4th degree (Table 5.5).
- systemic effects associated with burns depend on severity and extent and include:
  - severe anaemia
  - haemoglobinaemia and haemoglobinuria
  - hypernatraemia.
  - immediate hyperkalaemia
  - delayed diuretic hypokalaemia.
- electrolyte monitoring during the first 2 weeks post-wounding is paramount:
  - avoid serious electrolyte shifts and ensure optimal systemic management.
- high risk of developing life-threatening septicaemia:
  - systemic antibiotics are recommended.
- catabolic state follows burn injuries in horses:
  - high-calorie diet, and possibly anabolic steroids, are required.
- pulmonary damage by smoke inhalation is common:
  - direct damage to the airways and lung tissue by:
    - exposure to heat, smoke particulate matter and gaseous by-products.

TABLE 5.5 Classification of burns according to the degree of tissue damage	
CLASSIFICATION	**SIGNS**
First degree	Erythema, oedema, pain, desquamation of superficial layers
Second degree superficial	Fluid accumulation between stratum granulosum and basal cell layers (blister), moderate pain
Second degree deep	Oedema fluid at epidermal–dermal junction, epidermal necrosis, increase in WBCs at basal layer, eschar production, minimal pain
Third degree	Loss of epidermal and dermal elements, fluid and cell response at the margins and deeper tissue, eschar formation, lack of pain, shock, infection
Fourth degree	Carbonisation of tissue, deep tissue destruction as far down as muscle, bone, etc.

- ◦ bronchospasm and bronchoconstriction.
- ◦ carbon monoxide poisoning.
- ◦ pulmonary oedema
- ◦ acute respiratory distress
- ◦ pneumonia.
- ◦ treatment should involve:
  - ♦ supplemental moistened oxygen (if difficulty in breathing).
  - ♦ bronchodilators.
- overall therapy of burn patients has several stages, with different goals in mind depending on their severity.
- 1st stage (0–4 days):
  - ◦ systemic and respiratory stabilisation of the horse.
  - ◦ analgesia ◦ wound protection.
  - ◦ prevention of sepsis.
- 2nd stage (5–10 days):
  - ◦ progressive wound debridement.
  - ◦ wound protection.
  - ◦ systemic support.
  - ◦ prevention of sepsis.
- 3rd stage (10 days onwards):
  - ◦ wound repair ◦ protection.
  - ◦ restoration of epithelium.
- **First-degree burns**
  - ◦ cooled with ice or cold water.
  - ◦ prevent infection with a topical antibacterial preparation.
  - ◦ protect wound from further trauma although some may be left uncovered.
  - ◦ systemic analgesia.
- **Superficial second-degree burns**
  - ◦ characterised by formation of a blister (leave intact if possible).
  - ◦ blister bursts:
    - ♦ remove the tissue and clean, before protecting area with a bandage.
  - ◦ scabs may provide adequate initial protection to a wound:
    - ♦ if not disturbed or becomes infected.
  - ◦ non-adherent dressing, or petrolatum-, or antibiotic-impregnated gauze applied as a contact layer.
  - ◦ bandage changed as required depending on:
    - ♦ extent of wound ♦ location.
    - ♦ amount of exudate present.
    - ♦ possible bacterial contamination.
- **Deep second-degree burns and third-degree burns:**
  - ◦ require closed technique treatment (includes the use of an occlusive bandage):
    - ♦ not if infection, scab or more exudate.
  - ◦ **eschar technique** allows the wound to be protected by the eschar:
    - ♦ works best in small burnt areas.
    - ♦ not in large burns or areas where the burn may become traumatised.
    - ♦ wound is left open and trauma and/or infection may occur.
  - ◦ **semi-open** treatment involves the continuous application of moist bandages and antibacterial agents to the eschar:
    - ♦ moist dressings prevent heat and moisture loss – protects the eschar.
    - ♦ helps prevent bacterial contamination and infection.
    - ♦ frequent bandage changes allow frequent wound debridement:
      - – time-consuming
      - – controls amount of tissue removed.

Wounds involving synovial structures

Overview

- can be devastating and potentially life-threatening.
- infection within a synovial structure produces:
  - ◦ dramatic joint/sheath inflammatory response and marked synovial effusion.
  - ◦ effusion is easily detected except where joint/sheath not clearly palpable:
    - ♦ hip joint
    - ♦ periarticular cellulitis and severe oedema are present.
- **wounds in the vicinity of a joint or tendon sheath must be thoroughly investigated to rule out synovial involvement** (Fig. 5.31):
  - ◦ careful sterile digital exploration.
  - ◦ instil intrasynovial sterile lactated Ringer's solution to investigate communication with the wound.
  - ◦ failure to diagnose penetration promptly delays initial therapy and outcome.
- clinical pain level may be confusing in diagnosing synovial involvement:

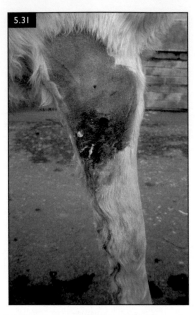

FIG. 5.31 This pony received a kick wound to the left lateral elbow region and penetration of the joint has occurred. Note the discharging synovial fluid.

- o particularly where the synovial structure is draining through the wound.
- centesis of the synovial structure at a remote location to the wound is essential:
  - o clinical judgement to assess the benefits and risks of centesis.
  - o severe surrounding cellulitis may cause contamination during centesis:
    - ◆ aggressive anti-inflammatory and antibiotic therapy.
    - ◆ centesis can be performed safely when the inflammation subsides.
- radiographic and ultrasonographic evaluation is recommended (Fig. 5.32):
  - o bone involvement or foreign bodies will affect treatment and prognosis.
- open synovial cavities are easier to detect and have better prognosis than puncture wounds.
- mixed bacterial populations are usually present:
  - o broad-spectrum antimicrobial therapy should be used.
  - o coliforms, *Streptococcus* spp. and *Staphylococcus* spp. are the most common.

- o local and systemic routes of administration.
- o aggressive long-term (around 6 weeks) therapy is often necessary.
- combined with anti-inflammatory drugs, analgesia, and synovial lavage or drainage.
- sterile bandage changes are required when a synovial structure has been penetrated.
- overall prognosis depends on prompt diagnosis and aggressive treatment.

Management of synovial sepsis in the adult

- confirmation of infectious synovial process:
  - o centesis
  - o cytological evaluation of synovial fluid.
  - o previous intra-articular steroid injection (up to 3 weeks) risk factor for sepsis:
    - ◆ may alter expected cytologic findings.
- synovial fluid examination:
  - o protein content.
  - o differential cell count
  - o cellular morphology:
    - ◆ 75% or more neutrophils or toxic changes to neutrophils are suspicious.
  - o serum and synovial levels of serum amyloid A (SAA):
    - ◆ confirm presence of sepsis and response to treatment.
- **any doubt pertaining to diagnosis, then assume infection is present and start treatment.**
- treatment of infectious synovitis is aimed at different areas:
  - o eliminating the causative organism(s).
  - o diminishing the inflammatory process and providing analgesia.
  - o restoring synovial homeostasis.
  - o rehabilitating the horse.

Eliminating the causative organism(s)

- bacterial elimination is achieved by antimicrobial therapy and synovial lavage.
- post-injection synovial infections commonly caused by *Staphylococcus aureus*.
- post-traumatic infections commonly caused by coliforms.

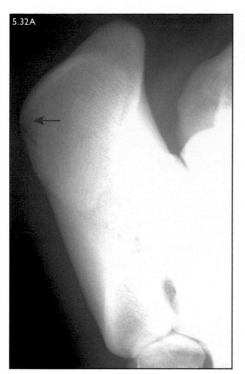

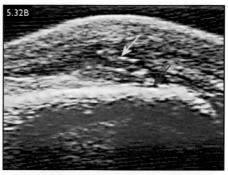

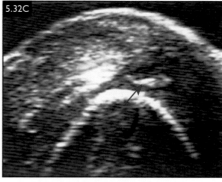

FIG. 5.32 (a) Dorso 45° lateral/plantaromedial oblique radiograph of the calcaneus of a horse that has received a kick wound to the proximal medial aspect of the point of the hock. Note the subtle lysis of the cortical surface of the calcaneus just below the point of the hock (arrow). **(b)** Longitudinal ultrasonogram of the wound area showing the superficial damage to the bone surface (red arrow), overlying soft-tissue proliferation, and mild distension of the inter-tendonous calcaneal bursa (yellow arrow). **(c)** Transverse ultrasonogram of the wound area showing a small separate fragment of bone to the lateral (left) side of the calcaneus (arrow).

- failure to obtain a positive culture happens commonly (up to 50% of cases):
  - educated guess as to what antibiotics may be most effective.
  - widely used antibiotic combinations:
    - penicillin or third-generation cephalosporin (e.g. ceftiofur, ceftazidime).
    - together with an aminoglycoside (e.g. amikacin).
  - systemic and local/regional administration (Table 5.6) are routine.
- synovial lavage is best performed by arthroscopy (Fig. 5.33):
  - through-and-through method using needles/drip sets in cost-sensitive cases.
  - refractory cases may require open joint drainage.

Reducing inflammation and providing analgesia

- adequate analgesia is very important:
  - permits early ambulation with positive effects on rehabilitation.
  - diminishes complications associated with pain and lack of movement:
    - impaction colic.
    - other limb overuse injuries.
    - adhesion formation in intrathecal infections.
- most common NSAIDs used are phenylbutazone, flunixin meglumine and ketoprofen:
  - phenylbutazone (2.2–4.4 mg/kg q24 h, depending on the response to therapy).
- severely painful cases, or where NSAIDs not indicated, opioids should be considered:

TABLE 5.6 Local and regional antimicrobial therapy options for infections of synovial structures

	INTRAOSSEUS	INTRAVENOUS	INTRASYNOVIAL
Method	Use 4.5 mm or 5.5 mm cannulated screw. Can be placed standing or under GA. Perfuse under sedation over 30–45 minutes.	Use a butterfly needle (25 gauge) or a catheter. Can be performed on saphenous, cephalic or abaxial sesamoid vessel.	Direct daily injection or use of constant rate infusion (CRI) pump.
Amikacin	Administer 1 mg/kg q24 h for 3–5 days diluted in 10 ml of total saline volume. Inject slowly over a period of 2–4 minutes.	Same as intraosseous.	Administer 125–250 mg by direct injection q24 h for 3–5 days. If used in a CRI pump, will deliver 10 ml (2,500 mg/450 kg) of amikacin per day.
Gentamicin	Administer 1–2 mg/kg q24 h for 3–5 days diluted in 10 ml of total saline volume. Inject slowly over a period of 2–4 minutes.	Same as intraosseous.	Administer 100–300 mg by direct injection q24 h for 3–5 days. If used in a CRI pump, will deliver 6 ml (600 mg/450 kg) of gentamicin per day.
Ceftiofur	Administer 1 mg/kg q24 h for 3–5 days diluted in 10 ml of total saline volume. Inject slowly over a period of 2–4 minutes.	Same as intraosseous.	Administer 150 mg by direct injection. If used in a CRI pump, make a 50% ceftiofur solution (25 mg/ml) and deliver 0.5 ml of solution per hour (300 mg/day).
Vancomycin	Administer 300 mg diluted to a maximum of 5 mg/ml and give at a speed of 2 ml/minute or 10 mg/minute, q24 h for 3–5 days.	Same as intraosseous.	Not recommended.
Advantages	Good concentration in synovial fluid, medullary cavity and tendons distal to perfusion site. Concentrations persist up to 24 hours and slightly longer compared with intravenous perfusion. Very well tolerated. Minimal side effects.	Very good concentration lasting over 24 hours.	Very good concentrations lasting over 24 hours.
Disadvantages	Need special equipment. Painful perfusion. Potential blemish over perforated bone. Avoid proximity to any tendon.	Very difficult in cases where cellulitis prevents identification of the vessel. Repeated injection necessary if catheter not placed. Catheter may kink or dislodge. Difficult to maintain in distal extremity. Sloughing of hoof possible if vascular damage occurs at the level of the distal extremity.	Daily injection needed. CRI device sometimes may not work properly.

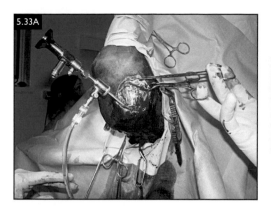
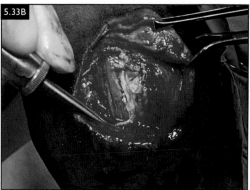

FIG. 5.33 This horse fell on the road when at exercise, lacerated the dorsal aspect of its right carpus, and subsequently had synovial fluid draining from the wound. Under GA and after surgical debridement/exploration of the wound, a defect into the dorsolateral aspect of the antebrachiocarpal joint was detected. This allowed an arthroscope to be inserted into the joint for exploration and lavage **(A)**. After the joint was lavaged, the tendon sheath of the extensor carpi radialis was also lavaged, as this had been involved in the trauma and was contaminated **(B)**.

- ○ epidural morphine (0.01–0.3 mg/kg) is useful in the hindlimbs.
- ○ systemic opioids alone, or in combination protocols, can help control pain:
  - ♦ butorphanol 0.02–0.1 mg/kg i/m or i/v q4–6 h on its own.
  - ♦ combination of ketamine 0.5–1 mg/kg and methadone 0.05 mg/kg i/m, q4–8 h.
- ○ Fentanyl patches (2–4 10 mg patches q72 h) (forelimbs):
  - ♦ provide good analgesia and comfort superior to NSAIDs.
- excessive use of analgesics can mask clinical deterioration or recrudescence of infection:
  - ○ clinician should titrate the analgesic protocol to the minimal dosage needed.

Restoring synovial homeostasis

- occurs over time once the inflammatory process has subsided:
  - ○ severe damage may lead to synovial homeostasis never being fully reached.
- intrasynovial hyaluronic acid provides anti-inflammatory effects and helps restore normality.
- passive rest followed by hand walking (post-acute inflammation) is also beneficial.
- systemic and/or oral glycosaminoglycans joint supplements may be useful.

Tendon lacerations

Flexor tendon

- commonly affect the superficial and/or deep digital flexor tendons (Fig. 5.34):
  - ○ rarely, other tendonous structures as well.
  - ○ characteristic appearance of a distal limb that has lost flexor support.
  - ○ where the suspensory ligament has also been severed or torn:
    - ♦ fetlock will be dramatically dropped (Fig. 5.35).
- palmar/plantar support lost/severely compromised:
  - ○ immediate support of the distal limb is mandatory:
    - ♦ failure to do so may predispose to hyperextension of the limb.
    - ♦ overstretching and damage to the palmar/plantar blood vessels.
- forelimb injuries may result from overreaching in racehorses:
  - ○ commonly associated with small wounds.
  - ○ wound size is not related to damage severity.
- flexor tendon lacerations may be within (intrathecal) or outside (extrathecal) tendon sheath:
  - ○ former are more difficult to manage due to a contaminated synovial cavity.

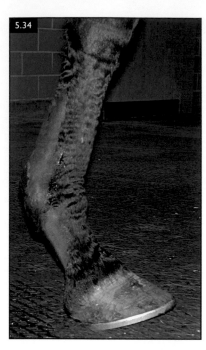

FIG. 5.34 The forelimb of a horse following a small laceration in the distal metacarpal area. Despite the small size of the laceration, loss of flexor tendon (deep digital flexor) support is obvious by the elevated position of the toe.

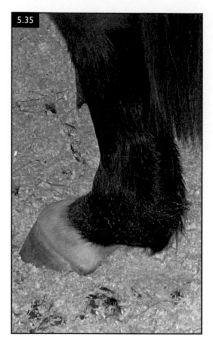

FIG. 5.35 Abnormal fetlock angle as a result of 'suspensory apparatus breakdown'. Note the almost horizontal position of the first phalanx.

- clinical management and prognosis vary according to:
  - location of the injury and which structure/s are involved.
- prompt and aggressive therapy is essential:
  - management of lacerated flexor tendons requires time and money.
  - many horses may not return to their previous level of athletic activity.
  - subsequent rehabilitation period often lasts an entire year.
- use of tenorrhaphy is controversial (Fig. 5.36):
  - advocated in cases where the tendon edges have not been severely damaged.
  - injury is intrathecal.
  - allows quicker gain in tensile strength during the first 12 weeks:
    - ◆ facilitates management of weight-bearing.
- lacerated tendons should be maintained in a cast for 8–12 weeks depending on severity of damage and response to treatment:
  - wounds requiring daily care should be placed in a bivalved cast or a bandage–cast combination.
- weight-bearing management and rehabilitation protocol for flexor tendon lacerations:
  - Weeks 1–6. Non-weight-bearing. Limb in a cast or Kimzey splint. Stall rest.
  - Weeks 7–9. As for previous 6 weeks or Patten shoe applied with heel elevation at approximately 30°. Limb maintained in a Robert Jones bandage. Stall rest.
  - Weeks 10–12. Patten shoe dropped at 15°. Stall rest.
  - Weeks 12–16. Patten shoe replaced by shoe with elevated heels at 12°. Stall rest.
  - Weeks 17–20. Heel elevation at 9°. Hand walking for 5 minutes once daily.
  - Weeks 21–24. Heel elevation at 9°. Hand walking for 10 minutes once daily.
  - Weeks 25–28. Heel elevation at 6°. Hand walking for 5 minutes once daily.

5

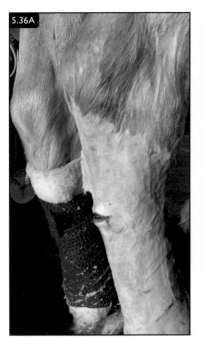
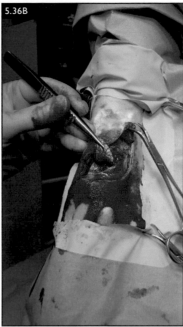

FIG. 5.36 There is a small laceration to the skin of the upper plantar cannon of the right hindlimb of this horse **(A)**. At surgical exploration of the wound, there is a partial laceration of the plantar surface of the superficial digital flexor tendon, which was debrided and repaired **(B)**.

- ○ Weeks 29–32. Heel elevation at 6°. Hand walking for 10 minutes once daily.
- ○ Weeks 33–40. Heel elevation at 3°. Hand walking for 5 minutes once daily with 5-minute increments every week until week 40.
- ○ Weeks 41–44. Heel elevation at 3°. Hand walking for 20 minutes once daily. Light trot for 5 minutes once daily.
- ○ Weeks 45–48. Normal shoe. Carry on exercise at 20 minutes walking exercise under saddle and 5 minutes trot on a lunge line with 5-minute weekly increments.
- ○ Weeks 49–52. Start light flat work.
- • ultrasonographic evidence of healing should be obtained every 8 weeks during the healing process.
- • Equestride™ fetlock support boot allows control of the progressive loading of the fetlock:
  - ○ can be used for rehabilitating flexor tendon lesions and lacerations (Fig. 5.37).

FIG. 5.37 An Equestride™ fetlock support boot.

Extensor tendon

- usually occur at the level of the cannon bone in either the fore- or hindlimbs.
- common (forelimb) or long (hindlimb) extensor tendon in the distal limb:
  - lack of a synovial membrane simplifies laceration management.
- tenorrhaphy is not necessary:
  - maintenance in a cast, bandage or splint is usually sufficient.
- general principles of wound care apply:
  - edges of the tendon can be debrided to encourage fibroplasia.
- initially, the horse is unable to extend the limb:
  - cast or splint is recommended during the healing process (6–8 weeks).
  - not uncommon to find the edges of the tendon 5–7.5 cm (2–3 inches) apart:
    - ♦ does not seem to affect the final outcome.
- most horses rehabilitate well and return to their previous level of activity.

Haematomas

- collection of free blood in the tissues usually caused by high-impact contusions:
  - kicks, self-inflicted during casting episodes, fence trapping or rope burns
  - damages blood vessels of different calibre and determines size of the haematoma.

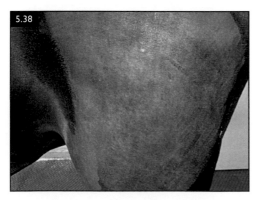

FIG. 5.38 A haematoma of the thigh region in a horse following a kick. Note the area of discolouration in the central aspect of the thigh. These injuries may potentially lead to compartment syndrome, with catastrophic consequences if not attended to promptly.

- usually do not result in skin breakage (Fig. 5.38).
- ideal environment for bacterial colonisation/proliferation, especially if large:
  - ♦ infection leads to abscess formation and tissue sloughing.
- severity of clinical signs and consequences depend on:
  - size, location and whether or not it is colonised by bacteria.
- **accumulation of fluid beneath a rigid fascial plane may result in the development of compartment syndrome:**
  - interrupts lymphatic drainage and blood supply to the surrounding tissues.
  - produces severe oedema, inflammation and tissue necrosis (see Fig. 5.2).
  - present with severe lameness, pain and mildly swollen and turgid affected region.
  - most common in the external aspect of the thigh, affecting the quadriceps region.
  - initial lack of skin breakage, and severe and non-specific signs, means haematomas leading to compartment syndrome are usually unnoticed.
  - **subsequent regional cellulitis may become a serious/life-threatening injury:**
    - ♦ horses should be treated promptly.
- haematomas should be investigated by ultrasound:
  - clinicians should exercise extreme caution if aspirating the affected region.
- treatment in the initial stages of a haematoma:
  - area should be iced to produce vasoconstriction and stop further bleeding.
  - after the initial 48 hours, followed with warm therapy to facilitate reabsorption.
  - aggressive anti-inflammatory therapy, broad-spectrum antibiotics and diuretics:
    - ♦ reduce the inflammation and oedema present.
    - ♦ prevent or treat a possible infection.

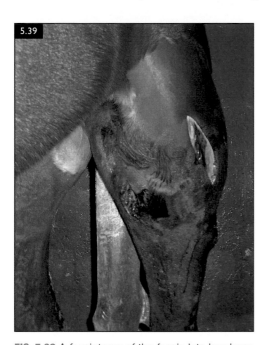

FIG. 5.39 A fasciotomy of the fascia lata has been performed in this case following a severe haematoma of the quadriceps musculature. Tissue necrosis had already occurred at the time of presentation and aggressive therapy was therefore deemed necessary.

- compartment syndrome is diagnosed:
  - area should be opened, and a fasciotomy performed (Fig. 5.39).
  - drains the tissues, releases the pressure and prevents tissue necrosis.
  - advanced stages of compartment syndrome, tissue necrosis and sloughing occur:
    - prognosis is grave even with aggressive therapy.
- following the resolution of a large haematoma:
  - may result in tissue fibrosis and calcification.
  - potentially leading to gait abnormalities.

Index